R Graphics

Chapman & Hall/CRC
Computer Science and Data Analysis Series

The interface between the computer and statistical sciences is increasing, as each discipline seeks to harness the power and resources of the other. This series aims to foster the integration between the computer sciences and statistical, numerical, and probabilistic methods by publishing a broad range of reference works, textbooks, and handbooks.

SERIES EDITORS
John Lafferty, Carnegie Mellon University
David Madigan, Rutgers University
Fionn Murtagh, Royal Holloway, University of London
Padhraic Smyth, University of California, Irvine

Proposals for the series should be sent directly to one of the series editors above, or submitted to:

Chapman & Hall/CRC
23-25 Blades Court
London SW15 2NU
UK

Published Titles

Bayesian Artificial Intelligence
Kevin B. Korb and Ann E. Nicholson

Pattern Recognition Algorithms for Data Mining
Sankar K. Pal and Pabitra Mitra

Exploratory Data Analysis with MATLAB®
Wendy L. Martinez and Angel R. Martinez

Clustering for Data Mining: A Data Recovery Approach
Boris Mirkin

Correspondence Analysis and Data Coding with Java and R
Fionn Murtagh

R Graphics
Paul Murrell

Computer Science and Data Analysis Series

R Graphics

Paul Murrell

The University of Auckland
New Zealand

Chapman & Hall/CRC
Taylor & Francis Group
Boca Raton   London   New York   Singapore

Published in 2006 by
Chapman & Hall/CRC
Taylor & Francis Group
6000 Broken Sound Parkway NW, Suite 300
Boca Raton, FL 33487-2742

International Standard Book Number-10: 1-58488-486-X (Hardcover)
International Standard Book Number-13: 978-1-58488-486-6 (Hardcover)
Library of Congress Card Number 2005046278

Library of Congress Cataloging-in-Publication Data

Murrell, Paul.
　R graphics / Paul Murrell.
　　　p. cm.
　Includes bibliographical references and index.
　ISBN 1-58488-486-X
　1. Computer graphics. 2. R (Computer program language) I. Title.

T385.M9 2005
006.6—dc22　　　　　　　　　　　　　　　　　　　　　　　　　　　2005046278

Taylor & Francis Group
is the Academic Division of T&F Informa plc.

Visit the Taylor & Francis Web site at
http://www.taylorandfrancis.com

and the CRC Press Web site at
http://www.crcpress.com

Preface

R is a popular open source software tool for statistical analysis and graphics. This book focuses on the very powerful graphics facilities that R provides for the production of publication-quality diagrams and plots.

What this book is about

This book describes the graphics system in R. The first chapter provides an overview of the R graphics facilities. There are many pictures that demonstrate the variety and complexity of plots and diagrams that can be produced using R. There is a description of the different output formats that R graphics can produce and there is a description of the overall organization of the R graphics facilities, so that the user has some idea of where to find a function for a particular purpose.

The most important feature of the R graphics setup is the existence of two distinct graphics systems within R: the traditional graphics system and the grid graphics system. Section 1.2.2 offers some advice on which system to use. Part I of this book is concerned with the traditional graphics system, which implements many of the "traditional" graphics facilities of the S language[11][5] (originally developed at Bell Laboratories and available in a commercial implementation as S-PLUS). The majority of R graphics functions available at the time of writing are based upon this system. The chapters in this part of the book describe how to work with the traditional graphics functions, with a particular emphasis on how to modify or add output to a plot to produce exactly the right final output. Chapter 2 describes the functions that are available to produce complete plots and Chapter 3 focuses on how to customize the details of plots, combine multiple plots, and add further output to plots.

Part II describes the grid graphics system, which is unique to R and is much more powerful than the traditional system. At the time of writing, there are fewer functions based on grid for producing complete plots, but there is more power to produce a wider range of final results. Most of the functions that produce complete plots using grid come from Deepayan Sarkar's lattice package, which implements Bill Cleveland's Trellis graphics system. This is described in Chapter 4. The remaining chapters describe how the grid system can be used to produce graphical scenes starting from a blank page. In particular, there is a discussion of how to develop new graphical functions that are easy for other people to use and build on.

Appendix A provides a very brief introduction to the R system in general and Appendix B discusses ways in which the traditional and grid graphics systems can be combined.

The main part of the book assumes a basic familiarity with the R language and environment. For more detailed information, the reader is directed to the home page of the R Project (the URL is given below), which has links to on-line documents and references to printed material.

There are a number of projects working on graphical user interfaces to R, but the common underlying method of interaction is via a command line. This book focuses on the production of graphical output by entering R code interactively at the command-line interface to R and writing code in scripts to load into R or to run as a batch job.

What this book is not about

This book does *not* contain discussions about which sort of plot is most appropriate for a particular sort of data, nor does it contain guidelines for correct graphical presentation. In fact, instructions are provided for producing some types of plots and graphical elements that are generally disapproved of, such as pie charts and cross-hatched fill patterns.

The information in this book is meant to be used to produce a plot once the format of the plot has been decided upon and to experiment with different ways of presenting a set of data. No plot types are deliberately excluded, partly because no plot type is all bad (e.g., a pie chart can be a very effective way to present a simple proportion) and partly because some graphical elements, such as cross-hatching, are sometimes required by a particular publisher.

The flexibility of R graphics encourages the user *not* to be constrained to thinking in terms of just the traditional types of plots. The aim of this book is to provide lots of useful tools and to describe how to use them. There are many other sources of information on graphical guidelines and recommended plot types, some of which are mentioned below.

Most introductory statistics text books will contain basic guidelines for selecting an appropriate type of plot. Examples of books that deal specifically with the construction of effective plots and are aimed at a general audience are "Creating More Effective Graphs" by Naomi Robbins[51] and Edward Tufte's "Visual Display of Quantitative Information"[60] and "Envisioning Information"[61]. For more technical discussions of these issues, see "Visualizing Data" and "Elements of Graphing Data" by Bill Cleveland[12][13], and "The Grammar of Graphics" by Leland Wilkinson[67].

For ideas on appropriate graphical displays for particular types of analysis or particular types of data, some starting points are "Data analysis and graphics using R" by John Maindonald and John Braun[37], "An R and S-Plus Companion to Applied Regression" by John Fox[20], "Statistical Analysis and Data Display" by Richard Heiberger and Burt Holland[29], and "Visualizing Categorical Data" by Michael Friendly[25].

This book is also *not* a complete reference to the R system. Appendix A provides a very brief introduction to R, but there are many freely-available documents that provide both introductory and in-depth explanations of the R system. The best place to start is the "Documentation" section on the home page of the R project web site (see "On the web" on page ix). Two examples of introductory texts are "Introductory Statistics with R" by Peter Dalgaard[18] and "Using R for Introductory Statistics" by John Verzani[65]; the standard advanced text is "Modern Applied Statistics with S" by Bill Venables and Brian Ripley[64].

Finally, this book does *not* describe in any detail the many graphics functions that are available in add-on packages for R that are *not* part of the standard R installation. This book only focuses on the graphics facilities that are distributed with R by default — in particular, functions in the grDevices, graphics, grid, and lattice packages. No attempt is made to enumerate all existing graphics functions for R or even to list all add-on packages that contain graphics functions; the list is very long and growing all the time. Except where specified, all add-on packages mentioned in this book are available from CRAN*, the main download site for R.

Differences with S-PLUS

The traditional graphics system in R is a reimplementation of the traditional graphics system in the original S language. This means that much of what is said about the traditional system in Part I of this book is also true for the traditional graphics in the commercial distribution of S, S-PLUS. However, there are some important differences between traditional R graphics and traditional S graphics, such as the specification of colors and line types by character strings, the concept of layouts for arranging plots, and the availability of mathematical annotation in text. These differences mean that graphics code written for R is not guaranteed to produce the same result (or even run) in S-PLUS. Furthermore, the grid graphics system described in Part II is not available in S-PLUS (just as the S-PLUS editable graphics are not available in R).

This book focuses on the graphics systems available in R so specific differences

---

*The Comprehensive R Archive Network; http://cran.r-project.org

with S-PLUS are not highlighted in the main text. However, much of what is said in Part I will also apply to traditional graphics in S-PLUS.

Who should read this book

This book should be of interest to a variety of R users. For people who are new to R, this book provides an overview of the graphics system, which is useful for understanding what to expect from R's graphics functions and how to modify or add to the output they produce. For this purpose, Chapter 1 and Chapter 2 are a good starting point from which to begin producing standard plots, but you will soon need to start dipping into Chapter 3 in order to fine tune your plots. It would also be worthwhile to take a look at Chapter 4 to see what Trellis plots can do.

For intermediate-level R users, this book provides all of the information necessary to perform sophisticated customizations of plots produced in R. As with many software applications, it is possible to work with R for years and remain unaware of important and useful features. This book will be useful in making users aware of the full scope of R graphics, and in providing a description of the correct model for working with R graphics. Sections 1.2, 1.3, and Chapters 3 and 4 should be read first. Chapters 5, 6, and 7 should be read by users interested in experimenting with novel graphical displays.

For advanced R users, this book contains vital information for producing coherent, reusable, and extensible graphics functions. Advanced users should pay particular attention to Part II.

Conventions used in this book

This book describes a large number of R functions and there are many code examples. Samples of code that could be entered interactively at the R command line are formatted as follows:

```
> 1:10
```

where the > denotes the R command-line prompt and everything else is what the user should enter. When an expression is longer than a single line it will look like this:

```
> plot(1:10, 1:10, col="blue", lty="dashed",
        axes=FALSE, type="l")
```

with the additional lines indented appropriately.

Often, the functions described in this book are used for the side-effect of producing graphical output, so the result of running a function is represented by a figure. In cases where the result of a function is a value that we might be interested in, the result will be shown below the code that produced it and will be formatted as follows:

```
[1]  1  2  3  4  5  6  7  8  9 10
```

In some places, an entirely new R function is defined. Such code would normally be entered into a script file and loaded into R in one step (rather than being entered at the command line), so the code for new R functions will be presented in a figure and formatted as follows:

```
1    myfun <- function(x, y) {
2      plot(x ,y)
3    }
```

with line numbers provided for easy reference to particular parts of the code from the main text.

When referring to a function within the main text, it will be formatted in a `typewriter` font and will have parentheses after the function name, e.g., `plot()`.

When referring to the arguments to a function or the values specified for the arguments, they will also be formatted in a `typewriter` font, but they will not have any parentheses at the end, e.g., `x`, `y`, or `col="red"`.

When referring to an S3 class, statements will be of the form: "the `"classname"` class," using a typewriter font with the class name in double-quotes. However, when referring to an object that is an instance of a class, statements will be of the form: "the `classname` object," using a typewriter font, but without the double-quotes around the class name.

On the web

There is a web site (URL below) with errata and links to pages of PNG versions of all figures from the book and the R code used to produce them.

```
http://www.stat.auckland.ac.nz/~paul/RGraphics/rgraphics.html
```

There is also an `RGraphics` package containing functions to produce the figures in this book and all functions, classes, and methods defined in the book (see especially Chapter 7). This package is available from CRAN (see the footnote on page vii).

Version information

Software development is an ongoing process and this book can only provide a snapshot of R's graphics facilities. The descriptions and code samples in this book are accurate for R version 2.1.0 and above. Apart from a couple of places, mostly in Chapter 7, code examples are also accurate for R version 2.0.1. In each of these cases, there is a footnote to highlight the difference and, if possible, to provide information about how to modify the code so that it will work in R version 2.0.1. Much of the content of Part I is also accurate for earlier versions of R, but specific areas of incompatibility are not indicated in the text.

A new "minor" version of R is released approximately every six months. The most up-to-date information on the most recent versions of R and grid are available in the on-line help pages and at the home pages for the R Project and the grid package:

```
http://www.R-project.org/
http://www.stat.auckland.ac.nz/~paul/grid/grid.html
```

Acknowledgements

R graphics could not exist without R itself, so the first thanks go to Ross Ihaka and Robert Gentleman for starting the whole thing. Thanks to the R Core Team in general for making R such a reliable, high-quality piece of software, and to the wider R community for making working with R so rewarding and enjoyable.

The traditional R graphics system owes most of its success and popularity to the excellence of the design of the original S graphics system. Most credit for the R-specific extensions to the traditional system is due to Ross Ihaka. For the grid system, I am almost entirely to blame.

With regard to this book in particular, I would like to thank John Chambers, Ross Ihaka, Duncan Murdoch, Stefano Iacus, Deepayan Sarkar, and the anonymous reviewers for valuable feedback on the manuscript.*

Last, and most, thank you Ju.

Auckland,   *Paul Murrell*
New Zealand,
May 2005

---

*This manuscript was generated on a Fedora Core 1 Linux system using the LATEX document preparation system, Friedrich Leisch's Sweave package, several of the GNU software tools, and of course R.

Contents

List of Figures   xv

List of Tables   xviii

1 An Introduction to R Graphics   **1**
  1.1   R graphics examples 3
      1.1.1   Standard plots 3
      1.1.2   Trellis plots 4
      1.1.3   Special-purpose plots 5
      1.1.4   General graphical scenes 5
  1.2   The organization of R graphics 16
      1.2.1   Types of graphics functions 17
      1.2.2   Traditional graphics versus grid graphics 18
  1.3   Graphical output formats 19
      1.3.1   Graphics devices 20
      1.3.2   Multiple pages of output 21
      1.3.3   Display lists 21

I   TRADITIONAL GRAPHICS   **23**

2 Simple Usage of Traditional Graphics   **25**
  2.1   The traditional graphics model 26
  2.2   Plots of one or two variables 27
      2.2.1   Arguments to graphics functions 32
      2.2.2   Standard arguments 34
  2.3   Plots of multiple variables 35
  2.4   Modern plots and specialized plots 38
  2.5   Interactive graphics 41

3 Customizing Traditional Graphics   **43**
  3.1   The traditional graphics model in more detail 44
      3.1.1   Plotting regions 44
      3.1.2   The traditional graphics state 48
  3.2   Controlling the appearance of plots 54
      3.2.1   Colors 55
      3.2.2   Lines 59

	3.2.3	Text	60
	3.2.4	Data symbols	68
	3.2.5	Axes	70
	3.2.6	Plotting regions	74
	3.2.7	Clipping	76
	3.2.8	Moving to a new plot	76
3.3	Arranging multiple plots		77
	3.3.1	Using the traditional graphics state	77
	3.3.2	Layouts	78
	3.3.3	The split-screen approach	82
3.4	Annotating plots		83
	3.4.1	Annotating the plot region	83
	3.4.2	Missing values and non-finite values	88
	3.4.3	Annotating the margins	89
	3.4.4	Legends	92
	3.4.5	Axes	94
	3.4.6	Mathematical formulae	97
	3.4.7	Coordinate systems	99
	3.4.8	Bitmap images	106
	3.4.9	Special cases	106
3.5	Creating new plots		114
	3.5.1	A simple plot from scratch	115
	3.5.2	A more complex plot from scratch	115
	3.5.3	Writing traditional graphics functions	118

II GRID GRAPHICS   123

4 Trellis Graphics: the Lattice Package   125

4.1	The lattice graphics model	126
	4.1.1	Lattice devices	128
4.2	Lattice plot types	130
	4.2.1	The `formula` argument and multipanel conditioning .	133
	4.2.2	A nontrivial example	135
4.3	Controlling the appearance of lattice plots	137
4.4	Arranging lattice plots	140
4.5	Annotating lattice plots	142
	4.5.1	Panel functions and strip functions	144
	4.5.2	Adding output to a lattice plot	147
4.6	Creating new lattice plots	147

5 The Grid Graphics Model   149

5.1	A brief overview of grid graphics	150
	5.1.1	A simple example	151
5.2	Graphical primitives	154

	5.2.1	Standard arguments	158
5.3		Coordinate systems	159
	5.3.1	Conversion functions	162
	5.3.2	Complex units	164
5.4		Controlling the appearance of output	166
	5.4.1	Specifying graphical parameter settings	169
	5.4.2	Vectorized graphical parameter settings	170
5.5		Viewports	173
	5.5.1	Pushing, popping, and navigating between viewports .	174
	5.5.2	Clipping to viewports	179
	5.5.3	Viewport lists, stacks, and trees	181
	5.5.4	Viewports as arguments to graphical primitives	183
	5.5.5	Graphical parameter settings in viewports	184
	5.5.6	Layouts	185
5.6		Missing values and non-finite values	190
5.7		Interactive graphics	191
5.8		Customizing lattice plots	191
	5.8.1	Adding grid output to lattice output	193
	5.8.2	Adding lattice output to grid output	194

6	**The Grid Graphics Object Model**		**199**
6.1		Working with graphical output	200
	6.1.1	Standard functions and arguments	201
6.2		Grob lists, trees, and paths	203
	6.2.1	Graphical parameter settings in gTrees	205
	6.2.2	Viewports as components of gTrees	206
	6.2.3	Searching for grobs	206
6.3		Working with graphical objects off-screen	207
	6.3.1	Capturing output	209
6.4		Placing and packing grobs in frames	210
	6.4.1	Placing and packing off-screen	213
6.5		Other details about grobs	213
	6.5.1	Calculating the sizes of grobs	213
	6.5.2	Editing graphical context	216
6.6		Saving and loading grid graphics	217
6.7		Working with lattice grobs	217

7	**Developing New Graphics Functions and Objects**		**221**
7.1		An example	222
	7.1.1	Modularity	223
7.2		Simple graphics functions	223
	7.2.1	Embedding graphical output	225
	7.2.2	Facilitating annotation	227
	7.2.3	Editing output	229
	7.2.4	Absolute versus relative sizes	230

7.3 Graphical objects 231
   7.3.1 Overview of creating a new graphical class 231
   7.3.2 Defining a new graphical class 232
   7.3.3 Validating grobs 234
   7.3.4 Drawing grobs 236
   7.3.5 Editing grobs 241
   7.3.6 Sizing grobs 245
   7.3.7 Pre-drawing and post-drawing 246
   7.3.8 Completing the example 249
   7.3.9 Reusing graphical elements 251
   7.3.10 Other details 253
7.4 Querying grid 263

A A Brief Introduction to R   **265**
A.1 Obtaining and installing R 265
A.2 An environment for statistical computing and graphics 265
   A.2.1 Batch processing 267
   A.2.2 Data types 268
   A.2.3 Variables 269
   A.2.4 Indexing 270
   A.2.5 Data structures 270
   A.2.6 Formulae 273
   A.2.7 Expressions 273
   A.2.8 Packages 274
   A.2.9 Accessing data sets 274
   A.2.10 Getting help 275
A.3 A programming language 275
   A.3.1 Debugging 276
A.4 An object-oriented language 277

B Combining Traditional Graphics and Grid Graphics   **279**
B.1 The `gridBase` package 279
   B.1.1 Annotating base graphics using grid 279
   B.1.2 Embedding base graphics plots in grid viewports . . . 280
   B.1.3 Problems and limitations 283

Bibliography   **287**

Function Index   **293**

Concept Index   **297**

List of Figures

1.1 A simple scatterplot 2
1.2 Some standard plots 7
1.3 A customized scatterplot 8
1.4 A Trellis dotplot 9
1.5 A map of New Zealand produced using R 10
1.6 Some polar-coordinate plots 11
1.7 A novel decision tree plot 12
1.8 A table-like plot 13
1.9 Didactic diagrams 14
1.10 A music score 15
1.11 A piece of clip art 15
1.12 The structure of the R graphics system 16

2.1 Four variations on a scatterplot 28
2.2 Plotting an `lm` object 30
2.3 Plotting an `agnes` object 31
2.4 Modifying default `barplot()` and `boxplot()` output 33
2.5 Standard arguments for high-level functions 36
2.6 Plotting three variables 37
2.7 Plotting multivariate data 39
2.8 Some modern and specialized plots 40

3.1 The plot regions in traditional graphics 45
3.2 Multiple figure regions in traditional graphics 46
3.3 The user coordinate system in the plot region 47
3.4 Figure margin coordinate systems 49
3.5 Outer margin coordinate systems 50
3.6 Predefined and custom line types 61
3.7 Line join and line ending styles 62
3.8 Alignment of text in the plot region 63
3.9 Font families and font faces 67
3.10 Data symbols available in R 69
3.11 Basic plot types 71
3.12 Different axis styles 72
3.13 Graphics state settings controlling plot regions 75
3.14 Some basic layouts 79

3.15 Some complex layouts 81
3.16 Annotating the plot region 84
3.17 More examples of annotating the plot region 87
3.18 Drawing polygons 89
3.19 Annotating the margins 91
3.20 Some simple legends 93
3.21 Customizing axes 95
3.22 Mathematical formulae in plots 98
3.23 Custom coordinate systems 100
3.24 Overlaying plots 102
3.25 Overlaying output 105
3.26 Adding a bitmap to a plot 107
3.27 Special-case annotations 109
3.28 A panel function example 111
3.29 Annotating a 3D surface 113
3.30 A back-to-back barplot 116
3.31 A graphics function template 121

4.1 A scatterplot using lattice 127
4.2 The result of modifying a lattice object 129
4.3 Plot types available in lattice 132
4.4 A lattice multipanel conditioning plot 134
4.5 A complex lattice plot 136
4.6 Some default lattice settings 138
4.7 Controlling the layout of lattice panels 141
4.8 Arranging multiple lattice plots 143
4.9 Annotating a lattice plot 145

5.1 A simple scatterplot using grid. 153
5.2 Primitive grid output 156
5.3 Drawing arrows 157
5.4 Drawing polygons 158
5.5 A demonstration of grid units 163
5.6 Graphical parameters for graphical primitives. 169
5.7 Recycling graphical parameters. 171
5.8 Recycling graphical parameters for polygons 172
5.9 A diagram of a simple viewport 174
5.10 Pushing a viewport 175
5.11 Pushing several viewports 176
5.12 Popping a viewport 177
5.13 Navigating between viewports 178
5.14 Clipping output in viewports 180
5.15 The inheritance of viewport graphical parameters 185
5.16 Layouts and viewports 187
5.17 Layouts and units 188

5.18 Nested layouts 190
5.19 Non-finite values for line-tos, polygons, and arrows 192
5.20 Controlling the size of lattice panels 193
5.21 Adding grid output to a lattice plot 195
5.22 Embedding a lattice plot within grid output 196

6.1   Modifying a circle grob 201
6.2   Editing grobs 203
6.3   The structure of a gTree 204
6.4   Editing a gTree 205
6.5   Using a gTree to group grobs 209
6.6   Packing grobs by hand 212
6.7   Calculating the size of a grob 215
6.8   Grob dimensions by reference 216
6.9   Editing the graphical context 217
6.10 Editing the grobs in a lattice plot 219

7.1   A plot of oceanographic data 222
7.2   A grid.imageFun() function 224
7.3   Output from the grid.imageFun() function 225
7.4   A grid.ozFun() function 226
7.5   Example output from grid.ozFun() 227
7.6   Annotating grid.ozFun() output 228
7.7   Editing grid.ozFun() output 230
7.8   An "imageGrob" class 233
7.9   Some validDetails() methods 235
7.10 An "ozGrob" class 238
7.11 An "ozImage" class 240
7.12 Some editDetails() methods 243
7.13 Editing an imageGrob 244
7.14 Low-level editing of an imageGrob 245
7.15 Helper functions for a "ribbonLegend" class 247
7.16 A "ribbonLegend" class 248
7.17 An "ozKey" class 250
7.18 A plot of temperature data 252
7.19 A splitString() function 255
7.20 Performing calculations before drawing 256
7.21 A "splitText" class 257
7.22 Drawing faces 260
7.23 Some face functions 261
7.24 Some face objects 262

B.1   Annotating a traditional plot with grid 281
B.2   Embedding a traditional plot within lattice output 284

List of Tables

1.1   Graphical output formats 20

3.1   High-level traditional graphics state settings 51
3.2   Low-level traditional graphics state settings 53
3.3   Read-only traditional graphics state settings 53
3.4   Functions to generate color sets 57
3.5   Font faces 66
3.6   Font families 66

4.1   Plotting functions in lattice 131

5.1   Graphical primitives in grid 155
5.2   Coordinate systems in grid 160
5.3   Graphical parameters in grid 167
5.4   Grid font faces 170

6.1   Functions for working with grobs 202

1

An Introduction to **R** Graphics

Chapter preview

This chapter provides the most basic information to get started producing plots in R. First of all, there is a three-line code example that demonstrates the fundamental steps involved in producing a plot. This is followed by a series of figures to demonstrate the range of images that R can produce. There is also a section on the organization of R graphics giving information on where to look for a particular function. The final section describes the different graphical output formats that R can produce and how to obtain a particular output format.

The following code provides a simple example of how to produce a plot using R (see Figure 1.1).

```
> plot(pressure)
> text(150, 600,
      "Pressure (mm Hg)\nversus\nTemperature (Celsius)")
```

The expression `plot(pressure)` produces a scatterplot of pressure versus temperature, including axes, labels, and a bounding rectangle.* The call to the `text()` function adds the label at the data location (`150, 600`) within the plot.

---

*The `pressure` data set, available in the `datasets` package, contains 19 recordings of the relationship between vapor pressure (in millimeters of mercury) and temperature (in degrees Celsius).

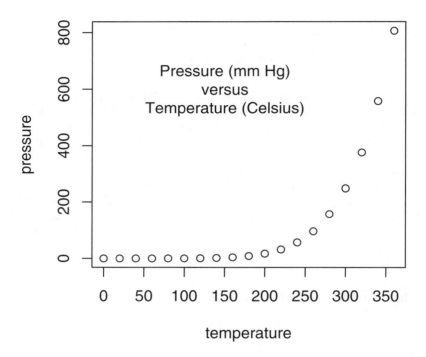

Figure 1.1

A simple scatterplot of vapor pressure of mercury as a function of temperature. The plot is produced from two simple R expressions: one expression to draw the basic plot, consisting of axes, data symbols, and bounding rectangle; and another expression to add the text label within the plot.

This example is basic R graphics in a nutshell. In order to produce graphical output, the user calls a series of graphics functions, each of which produces either a complete plot, or adds some output to an existing plot. R graphics follows a "painters model," which means that graphics output occurs in steps, with later output obscuring any previous output that it overlaps.

There are very many graphical functions provided by R and the add-on packages for R, so before describing individual functions, Section 1.1 demonstrates the variety of results that can be achieved using R graphics. This should provide some idea of what users can expect to be able to achieve with R graphics.

Section 1.2 gives an overview of how the graphics functions in R are organized. This should provide users with some basic ideas of where to look for a function to do a specific task. Section 1.3 describes the set of functions involved with the selection of a particular graphical output format. By the end of this chapter, the reader will be in a position to start understanding in more detail the core R functions that produce graphical output.

1.1 R graphics examples

This section provides an introduction to R graphics by way of a series of examples. None of the code used to produce these images is shown, but it is available from the web site for this book. The aim for now is simply to provide an overall impression of the range of graphical images that can be produced using R. The figures are described over the next few pages and the images themselves are all collected together on pages 7 to 15.

1.1.1 Standard plots

R provides the usual range of standard statistical plots, including scatterplots, boxplots, histograms, barplots, piecharts, and basic 3D plots. Figure 1.2 shows some examples.*

In R, these basic plot types can be produced by a single function call (e.g.,

---

*The barplot makes use of data on death rates in the state of Virginia for different age groups and population groups, available as the VADeaths data set in the datasets package. The boxplot example makes use of data on the effect of vitamin C on tooth growth in guinea pigs, available as the ToothGrowth data set, also from the datasets package. These and many other data sets distributed with R were obtained from "Interactive Data Analysis" by Don McNeil[40] rather than directly from the original source.

`pie(pie.sales)` will produce a piechart), but plots can also be considered merely as starting points for producing more complex images. For example, in the scatterplot in Figure 1.2, a text label has been added within the body of the plot (in this case to show a subject identification number) and a secondary y-axis has been added on the right-hand side of the plot. Similarly, in the histogram, lines have been added to show a theoretical normal distribution for comparison with the observed data. In the barplot, labels have been added to the elements of the bars to quantify the contribution of each element to the total bar and, in the boxplot, a legend has been added to distinguish between the two data sets that have been plotted.

This ability to add several graphical elements together to create the final result is a fundamental feature of R graphics. The flexibility that this allows is demonstrated in Figure 1.3, which illustrates the estimation of the original number of vessels based on broken fragments gathered at an archaeological site: a measure of "completeness" is obtained from the fragments at the site; a theoretical relationship is used to produce an estimated range of "sampling fraction" from the observed completeness; and another theoretical relationship dictates the original number of vessels from a sampling fraction[19]. This plot is based on a simple scatterplot, but requires the addition of many extra lines, polygons, and pieces of text, and the use of multiple overlapping coordinate systems to produce the final result.

For more information on the R functions that produce these standard plots, see Chapter 2. Chapter 3 describes the various ways that further output can be added to a plot.

1.1.2 Trellis plots

In addition to the traditional statistical plots, R provides an implementation of Trellis plots[6] via the package `lattice`[54] by Deepayan Sarkar. Trellis plots embody a number of design principles proposed by Bill Cleveland[12][13] that are aimed at ensuring accurate and faithful communication of information via statistical plots. These principles are evident in a number of new plot types in Trellis and in the default choice of colors, symbol shapes, and line styles provided by Trellis plots. Furthermore, Trellis plots provide a feature known as "multi-panel conditioning," which creates multiple plots by splitting the data being plotted according to the levels of other variables.

Figure 1.4 shows an example of a Trellis plot. The data are yields of several different varieties of barley at six sites, over two years. The plot consists of six "panels," one for each site. Each panel consists of a dotplot showing yield for each site with different symbols used to distinguish different years, and a "strip" showing the name of the site.

For more information on the Trellis system and how to produce Trellis plots using the lattice package, see Chapter 4.

1.1.3   Special-purpose plots

As well as providing a wide variety of functions that produce complete plots, R provides a set of functions for producing graphical output primitives, such as lines, text, rectangles, and polygons. This makes it possible for users to write their own functions to create plots that occur in more specialized areas. There are many examples of special-purpose plots in add-on packages for R. For example, Figure 1.5 shows a map of New Zealand produced using R and the add-on packages maps[7] and mapproj[39].

R graphics works mostly in rectangular Cartesian coordinates, but functions have been written to display data in other coordinate systems. Figure 1.6 shows three plots based on polar coordinates. The top-left image was produced using the stars() function. Such star plots are useful for representing data where many variables have been measured on a relatively small number of subjects. The top-right image was produced using customized code by Karsten Bjerre and the bottom-left image was produced using the rose.diag() function from the CircStats package[36]. Plots such as these are useful for presenting geographic, or compass-based data. The bottom-right image in Figure 1.6 is a ternary plot producing using ternaryplot() from the vcd package[41]. A ternary plot can be used to plot categorical data where there are exactly three levels.

In some cases, researchers are inspired to produce a totally new type of plot for their data. R is not only a good platform for experimenting with novel plots, but it is also a good way to deliver new plotting techniques to other researchers. Figure 1.7 shows a novel display for decision trees, visualizing the distribution of the dependent variable in each terminal node[30].

For more information on how to generate a plot starting from an empty page with traditional graphics functions, see Chapter 3. The grid package provides even more power and flexibility for producing customized graphical output (see Chapters 5 and 6), especially for the purpose of producing functions for others to use (see Chapter 7).

1.1.4   General graphical scenes

The generality and flexibility of R graphics makes it possible to produce graphical images that go beyond what is normally considered to be statistical graphics, although the information presented can usually be thought of as data of some kind. A good mainstream example is the ability to embed tabular arrangements of text as graphical elements within a plot as in Figure 1.8. This is a standard way of presenting the results of a meta-analysis. Figure 1.12 and Figure 3.6 provide other examples of tabular graphical output produced by R.*

R has also been used to produce figures that help to visualize important concepts or teaching points. Figure 1.9 shows two examples that provide a geometric representation of extensions to F-tests (provided by Arden Miller[42]). A more unusual example of a general diagram is provided by the musical score in Figure 1.10 (provided by Steven Miller). R graphics can even be used like a general-purpose painting program to produce "clip art" as shown by Figure 1.11. These examples tend to require more effort to achieve the final result as they cannot be produced from a single function call. However, R's graphics facilities, especially those provided by the grid system (Chapters 5 and 6), provide a great deal of support for composing arbitrary images like these.

These examples present only a tiny taste of what R graphics (and clever and enthusiastic users) can do. They highlight the usefulness of R graphics not only for producing what are considered to be standard plot types (for little effort), but also for providing tools to produce final images that are well beyond the standard plot types (including going beyond the boundaries of what is normally considered statistical graphics).

---

*All of the figures in this book, apart from the figures in Chapter 7 that only contain R code, were produced using R.

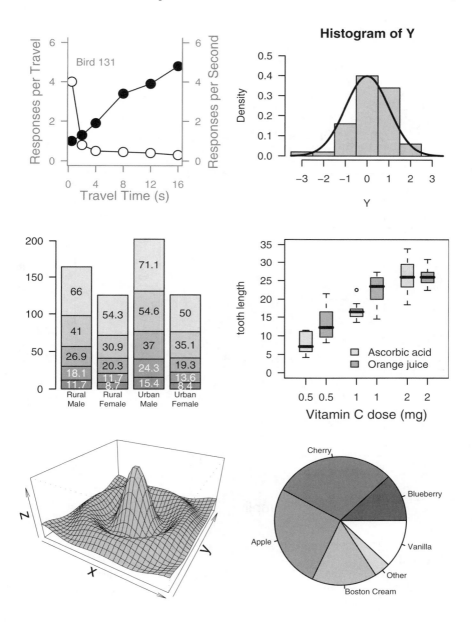

Figure 1.2

Some standard plots produced using R: (from left-to-right and top-to-bottom) a scatterplot, a histogram, a barplot, a boxplot, a 3D surface, and a piechart. In the first four cases, the basic plot type has been augmented by adding additional labels, lines, and axes. (The boxplot is adapted from an idea by Roger Bivand.)

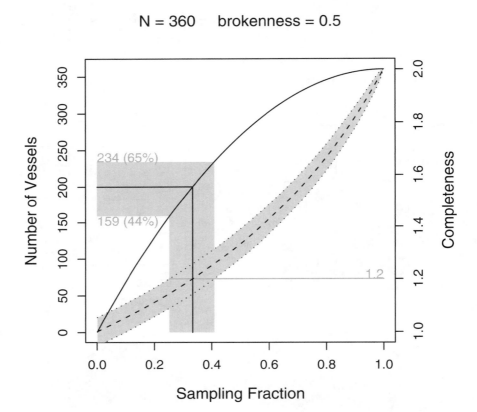

Figure 1.3
A customized scatterplot produced using R. This is created by starting with a simple scatterplot and augmenting it by adding an additional y-axis and several additional sets of lines, polygons, and text labels.

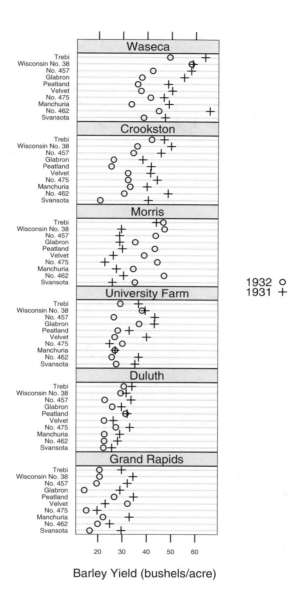

Figure 1.4

A Trellis dotplot produced using R. The relationship between the yield of barley and species of barley is presented, with a separate dotplot for different experimental sites and different plotting symbols for data gathered in different years. This is a small modification of Figure 1.1 from Bill Cleveland's "Visualizing Data" (reproduced with permission from Hobart Press).

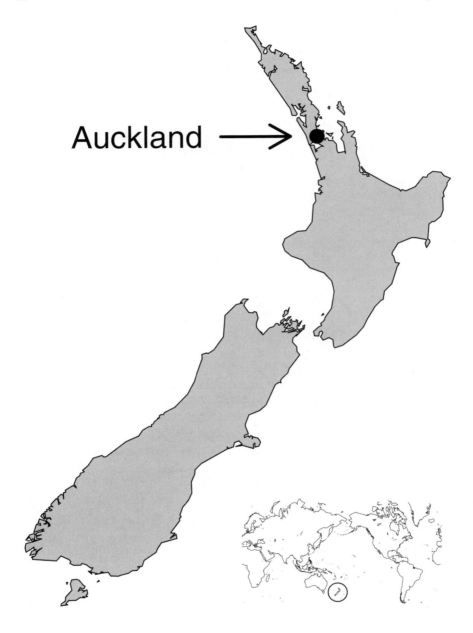

Figure 1.5

A map of New Zealand produced using R, Ray Brownrigg's `maps` package, and
Thomas Minka's `mapproj` package. The map (of New Zealand) is drawn as a series of polygons, and then text, an arrow, and a data point have been added to
indicate the location of Auckland, the birthplace of R. A separate world map has
been drawn in the bottom-right corner, with a circle to help people locate New
Zealand.

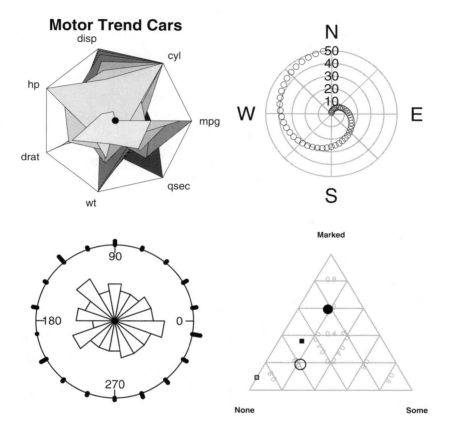

Figure 1.6
Some polar-coordinate plots produced using R (top-left), the CircStats package by Ulric Lund and Claudio Agostinelli (top-right), and code submitted to the R-help mailing list by Karsten Bjerre (bottom-left). The plot at bottom-right is a ternary plot produced using the vcd package (by David Meyer, Achim Zeileis, Alexandros Karatzoglou, and Kurt Hornik)

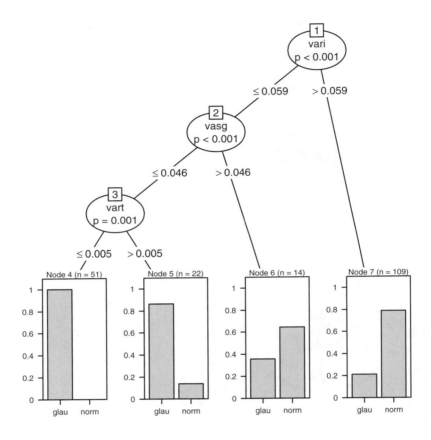

Figure 1.7
A novel decision tree plot, visualizing the distribution of the dependent variable
in each terminal node. From code under development by Torsten Hothorn, Kurt
Hornik, Achim Zeileis, and Friedrich Leisch and planned to appear on CRAN as the
package party.

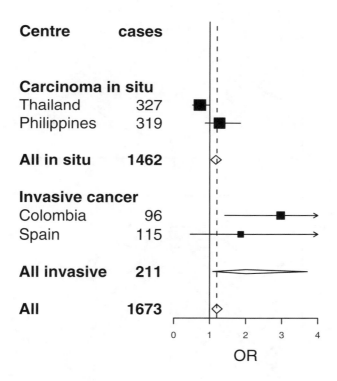

Figure 1.8
A table-like plot produced using R. This is a typical presentation of the results from a meta-analysis. The original motivation and data were provided by Martyn Plummer[48].

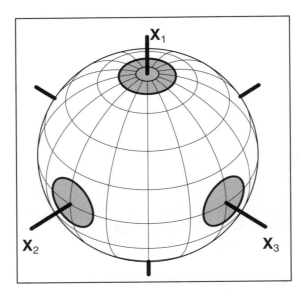

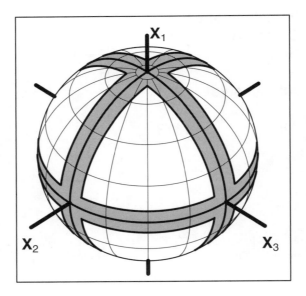

Figure 1.9
Didactic diagrams produced using R and functions provided by Arden Miller. The
figures show a geometric representation of extensions to F-tests.

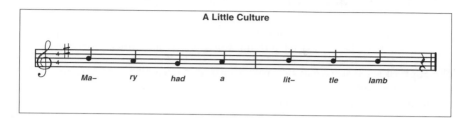

Figure 1.10

A music score produced using R (code by Steven Miller).

Figure 1.11

A piece of clip art produced using R.

1.2   The organization of **R** graphics

This section briefly describes how R's graphics functions are organized so that
the user knows where to start looking for a particular function.

The R graphics system can be broken into four distinct levels: graphics packages; graphics systems; a graphics engine, including standard graphics devices;
and graphics device packages (see Figure 1.12).

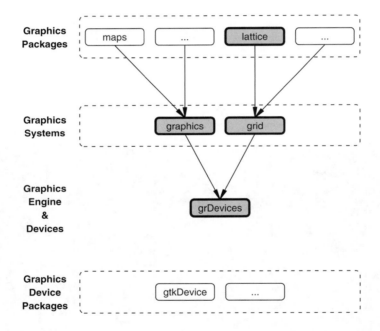

Figure 1.12
The structure of the R graphics system showing the main packages that provide
graphics functions in R. Arrows indicate where one package builds on the functions
in another package. The packages described in this book are highlighted with thicker
borders and grey backgrounds.

The core R graphics functionality described in this book is provided by the graphics engine and the two graphics systems, traditional graphics and grid. The graphics engine consists of functions in the **grDevices** package and provides fundamental support for handling such things as colors and fonts (see Section 3.2), and graphics devices for producing output in different graphics formats (see Section 1.3).

The traditional graphics system consists of functions in the **graphics** package and is described in Part I. The grid graphics system consists of functions in the **grid** package and is described in Part II.

There are many other graphics functions provided in add-on graphics packages, which build on the functions in the graphics systems. Only one such package, the **lattice** package, is described in any detail in this book. The **lattice** package builds on the grid system to provide Trellis plots (see Chapter 4).

There are also add-on graphics device packages that provide additional graphical output formats.

1.2.1 Types of graphics functions

Functions in the graphics systems and graphics packages can be broken down into three main types: *high-level* functions that produce complete plots; *low-level* functions that add further output to an existing plot; and functions for working interactively with graphical output.

The traditional system, or graphics packages built on top of it, provide the majority of the high-level functions currently available in R. The most significant exception is the lattice package (see Chapter 4), which provides complete plots based on the grid system.

Both the traditional and grid systems provide many low-level graphics functions, and grid also provides functions for interacting with graphical output (editing, extracting, deleting parts of an image).

Most functions in graphics packages produce complete plots and typically offer specialized plots for a specific sort of analysis or a specific field of study. For example: the **hexbin** package[10] from the BioConductor project has functions for producing hexagonal binning plots for visualizing large amounts of data; the **maps** package[7] provides functions for visualizing geographic data (see, for example, Figure 1.5); and the package **scatterplot3d**[35] produces a variety of 3-dimensional plots. If there is a need for a particular sort of plot, there is a reasonable chance that someone has already written a function to do it. For example, a common request on the **R-help** mailing list is for a way to add error bars to scatterplots or barplots and this can be achieved via the functions `plotCI()` from the `gplots` package in the `gregmisc` bundle or the `errbar()` function from the `Hmisc` package. There are some search facilities linked off the main R home page web site to help to find a particular function for a particular purpose (also see Section A.2.10).

While there is no detailed discussion of the high-level graphics functions in graphics packages other than lattice, the general comments in Chapter 2 concerning the behavior of high-level functions in the traditional graphics system will often apply as well to high-level graphics functions in graphics packages built on the traditional system.

1.2.2   Traditional graphics versus grid graphics

The existence of two distinct graphics systems in R raises the issue of when to use each system.

For the purpose of producing complete plots from a single function call, which graphics system to use will largely depend on what type of plot is required. The choice of graphics system is largely irrelevant if no further output needs to be added to the plot.

If it is necessary to add further output to a plot, the most important thing to know is which graphics system was used to produce the original plot. In general, the same graphics system should be used to add further output (though see Appendix B for ways around this).

In some cases, the same sort of plot can be produced by both lattice and traditional functions. The lattice versions offer more flexibility for adding further output and for interacting with the plot, plus Trellis plots have a better design in terms of visually decoding the information in the plot.

For producing graphical scenes starting from a blank page, the grid system offers the benefit of a much wider range of possibilities, at the cost of a having to learn a few additional concepts.

For the purpose of writing new graphical functions for others to use, grid again provides better support for producing more general output that can be combined with other output more easily. Grid also provides more possibilities for interaction.

1.3 Graphical output formats

At the start of this chapter (page 1), there is a simple example of the sort of R expressions that are required to produce a plot. When using R interactively, the result is a plot drawn on screen. However, it is also possible to produce a file that contains the plot, for example, as a PostScript document. This section describes how to control the format in which a plot is produced.

R graphics output can be produced in a wide variety of graphical formats. In R's terminology, output is directed to a particular output *device* and that dictates the output format that will be produced. A device must be created or "opened" in order to receive graphical output and, for devices that create a file on disk, the device must also be closed in order to complete the output. For example, for producing PostScript output, R has a function `postscript()` that opens a file to receive PostScript commands. Graphical output sent to this device is recorded by writing PostScript commands into the file. The function `dev.off()` closes a device.

The following code shows how to produce a simple scatterplot in PostScript format. The output is stored in a file called `myplot.ps`:

```
> postscript(file="myplot.ps")
> plot(pressure)
> dev.off()
```

To produce the same output in PNG format (in a file called `myplot.png`), the code simply becomes:

```
> png(file="myplot.png")
> plot(pressure)
> dev.off()
```

When working in an interactive session, output is often produced, at least initially, on the screen. When R is installed, an appropriate screen format is selected as the default device and this default device is opened automatically the first time that any graphical output occurs. For example, on the various Unix systems, the default device is an X11 window so the first time a graphics function gets called, a window is created to draw the output on screen. The user can control the format of the default device using the `options()` function.

Table 1.1
Graphics formats that R supports and the functions that open
an appropriate graphics device

Device Function	Graphical Format
Screen/GUI Devices	
x11() or X11()	X Window window
windows()	Microsoft Windows window
quartz()	Mac OS X Quartz window
File Devices	
postscript()	Adobe PostScript file
pdf()	Adobe PDF file
pictex()	LaTeX PicTeX file
xfig()	XFIG file
bitmap()	GhostScript conversion to file
png()	PNG bitmap file
jpeg()	JPEG bitmap file
(Windows only)	
win.metafile()	Windows Metafile file
bmp()	Windows BMP file
Devices provided by add-on packages	
devGTK()	GTK window (gtkDevice)
devJava()	Java Swing window (RJavaDevice)
devSVG()	SVG file (RSvgDevice)

1.3.1 Graphics devices

Table 1.1 gives a full list of functions that open devices and the output formats
that they correspond to.

All of these functions provide several arguments to allow the user to specify
things such as the physical size of the window or document being created. The
documentation for individual functions should be consulted for descriptions
of these arguments.

It is possible to have more than one device open at the same time, but only
one device is currently "active" and all graphics output is sent to that device.

If multiple devices are open, there are functions to control which device is
active. The list of open devices can be obtained using dev.list(). This gives
the name (the device format) and number for each open device. The function
dev.cur() returns this information only for the currently active device. The
dev.set() function can be used to make a device active, by specifying the

appropriate device number and the functions `dev.next()` and `dev.prev()` can be used to make the next/previous device on the device list the active device.

All open devices can be closed at once using the function `graphics.off()`. When an R session ends, all open devices are closed automatically.

1.3.2 Multiple pages of output

For a screen device, starting a new page involves clearing the window before producing more output. On Windows there is a facility for returning to previous screens of output (see the "History" menu, which is available when a graphics window has focus), but on most screen devices, the output of previous pages is lost.

For file devices, the output format dictates whether multiple pages are supported. For example, PostScript and PDF allow multiple pages, but PNG does not. It is usually possible, especially for devices that do not support multiple pages of output, to specify that each page of output produces a separate file. This is achieved by specifying the argument `onefile=FALSE` when opening a device and specifying a pattern for the file name like `file="myplot%03d"` so that the `%03d` is replaced by a three-digit number (padded with zeroes) indicating the "page number" for each file that is created.

1.3.3 Display lists

R maintains a *display list* for each open device, which is a record of the output on the current page of a device. This is used to redraw the output when a device is resized and can also be used to copy output from one device to another.

The function `dev.copy()` copies all output from the active device to another device. The copy may be distorted if the aspect ratio of the destination device — the ratio of the physical height and width of the device — is not the same as the aspect ratio of the active device. The function `dev.copy2eps()` is similar to `dev.copy()`, but it preserves the aspect ratio of the copy and creates a file in EPS (Encapsulated PostScript) format that is ideal for embedding in other documents (e.g., a LaTeX document). The `dev2bitmap()` function is similar in that it also tries to preserve the aspect ratio of the image, but it produces one of the output formats available via the `bitmap()` device.

The function `dev.print()` attempts to print the output on the active device. By default, this involves making a PostScript copy and then invoking the print command given by `options("printcmd")`.

The display list can consume a reasonable amount of memory if a plot is particularly complex or if there are very many devices open at the same time. For this reason it is possible to disable the display list, by typing the expression `dev.control(displaylist="inhibit")`. If the display list is disabled, output will not be redrawn when a device is resized, and output cannot be copied between devices.

Chapter summary

R graphics can produce a wide variety of graphical output, including (but not limited to) many different kinds of statistical plots, and the output can be produced in a wide variety of formats. Graphical output is produced by calling functions that either draw a complete plot or add further output to an existing plot.

There are two main graphics systems in R: a traditional system similar to the original S graphics system and a newer grid system that is unique to R. Additional graphics functionality is provided by a large number of add-on packages that build on these graphics systems.

Part I

TRADITIONAL GRAPHICS

2

Simple Usage of Traditional Graphics

Chapter preview

This chapter introduces the main *high-level* plotting functions in the traditional graphics system. These are the functions used to produce complete plots such as scatterplots, histograms, and boxplots. This chapter describes the names of the standard plotting functions, the standard ways to call these functions, and some of the standard arguments that can be used to vary the appearance of the plots. Some of this information is also applicable to high-level plotting functions in other add-on packages.

The aim of this chapter is to provide an idea of the range of functions that are available in the traditional graphics system, to point the user toward the most important ones, and introduce the standard approach to using them.

The graphics functions that make up the traditional graphics system are provided in an add-on package called **graphics**, which is automatically loaded in a standard installation of R. In a non-standard installation, it may be necessary to make the following call in order to access traditional graphics functions (if the **graphics** package is already loaded, this will not do any harm).

```
> library(graphics)
```

This chapter mentions all of the high-level graphics functions in the **graphics** package, but does not describe all possible uses of these functions. For detailed information on the behavior of individual functions the user should consult the individual help pages using the **help()** function (or **help.start()** for a web-browser interface). For example, the following code shows the help page for the `barplot()` function.

```
> help(barplot)
```

Another useful way of learning about a graphics function is to use the `example()` function. This runs the code in the "Examples" section of the help page for a function. The following code runs the examples for `barplot()`.

```
> par(ask=TRUE)
> example(barplot)
```

The `par(ask=TRUE)` is important to ensure that the user is prompted before each new page; without it the examples tend to flash by too fast for them to be viewed properly.

2.1   The traditional graphics model

As described at the start of Chapter 1, a plot is created in traditional graphics by first calling a high-level function that creates a complete plot, then calling low-level functions to add more output if necessary.

Traditional graphics functions always produce output on the current device (see Section 1.3.1 for information on devices and selecting a current device when more than one device is open). There is also the concept of a "current plot," and all low-level functions add output to the current plot. If there is only one plot per page, then a high-level function starts a new plot on a new page. There may be multiple plots on a page (see Section 3.3), and in this case a high-level function starts the next plot on the same page, only starting a new page when the number of plots per page is exceeded.

The main persistent record of graphical output is the device output — a window on screen or a file on disk. The only way to edit graphical output is to modify and rerun the original R code, or to produce output in a format that can be edited using third-party software (e.g., the output from an `xfig()` device can be edited using the `xfig` program; on Windows, the metafile format can be edited by a number of different programs).

2.2 Plots of one or two variables

The traditional graphics system provides a standard set of basic plot types. The `plot()` function produces scatterplots, the `barplot()` function produces barplots, `hist()` produces histograms, `boxplot()` produces boxplots, and `pie()` produces piecharts (see Figure 1.2 for example output).

R does not make a major distinction between, for example, scatterplots that only plot data symbols at each (x, y) location and scatterplots that draw straight lines connecting the (x, y) locations (line plots). These are just variations on the basic scatterplot, controlled by a `type` argument. This is demonstrated by the following code, which produces four different plots by varying the value of the `type` argument (see Figure 2.1).

```
> y <- rnorm(20)
> plot(y, type="p")
> plot(y, type="l")
> plot(y, type="b")
> plot(y, type="h")
```

R also does not make a distinction between a plot of a single set of data and a plot containing multiple series of data. Additional data series can be added to a plot using low-level functions such as `points()` and `lines()` (see Section 3.4.1; also see the function `matplot()` below).

The first argument to these high-level functions is the data to plot, but there is a reasonable amount of flexibility in the way that the data can be specified. For example, each of the following calls to `plot()` can be used to produce the scatterplot in Figure 1.1 (with small variations in the axis labels). In the first case, all of the data to plot are specified in a single data frame. In the second case, separate x and y variables are specified. In the third case, the data to plot are specified as a formula.

```
> plot(pressure)
> plot(pressure$temperature, pressure$pressure)
> plot(pressure ~ temperature, data=pressure)
```

All of the basic plotting functions in the traditional graphics system are generic (see Section A.4). One consequence of this has just been described — there are several ways to specify the data to plot — but this also means that in some cases the plot that the functions produce depends on the type of arguments

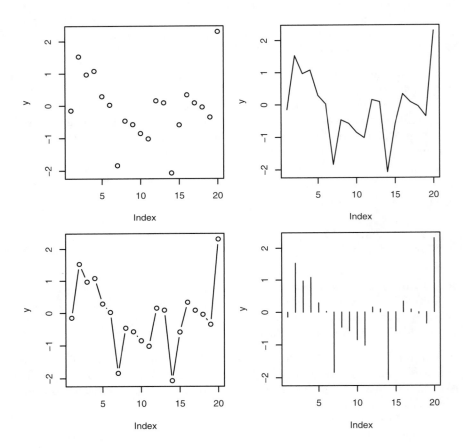

Figure 2.1
Four variations on a scatterplot. In each case, the plot is produced by a call to
the plot() function with the same data; all that changes is the value of the type
argument.

passed to the functions. This is most relevant to the `plot()` function, which, for example, will produce boxplots if the `x` variable is a factor. Another example is shown in the code below. Here an `lm` object is created from a call to the `lm()` function. When this object is passed to the `plot()` function, the special plot method for `lm` objects produces several regression diagnostic plots (see Figure 2.2).*

```
> lm.SR <- lm(sr ~ pop15 + pop75 + dpi + ddpi,
          data = LifeCycleSavings)
> plot(lm.SR)
```

In many cases, add-on graphics packages provide new plots by defining a new method for the `plot()` function. For example, the **cluster** package[52] provides a `plot()` method for plotting the result of an agglomerative hierarchical clustering procedure[32][53][56] (an **agnes** object). This method produces a special "bannerplot" and a dendrogram from the data (see the following code and Figure 2.3).† The first five expressions are just setting up the data; the last two expressions create an **agnes** object and then plot it.

```
> subset <- sample(1:150, 20)
> cS <- as.character(Sp <- iris$Species[subset])
> cS[Sp == "setosa"] <- "S"
> cS[Sp == "versicolor"] <- "V"
> cS[Sp == "virginica"] <- "g"
> ai <- agnes(iris[subset, 1:4])
> plot(ai, labels = cS)
```

The `matplot()` function is not a `plot()` method, but it is specifically designed to work like `plot()` with `x` or `y` given as matrices. This function is a convenient way to plot multiple data series on a single scatterplot. Different data series are automatically distinguished by using different data symbols and colors.

In addition to the very traditional set of plots, there is a function for producing scatterplots of a single variable, `stripchart()`, a function for drawing curves representing a mathematical function, `curve()`, and a function for producing a character-based stem-and-leaf plot, `stem()`.

---

*The data used in this example are measures relating to the savings ratio (aggregate personal saving divided by disposable income) averaged over the period 1960-1970 for 50 countries, available as the data set `LifeCycleSavings` in the **datasets** package.

†The data used in this example are the famous iris data data set giving measurements of physical dimensions of three species of iris, available as the `iris` data set in the **datasets** package.

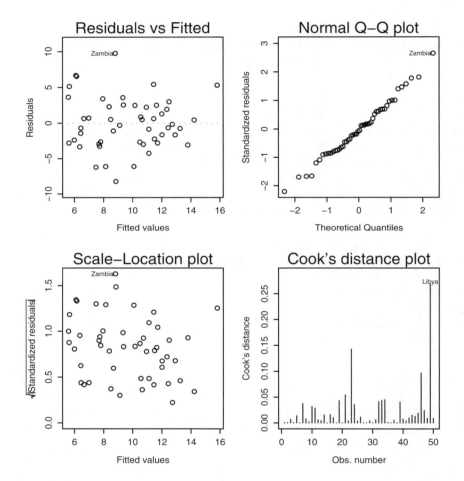

Figure 2.2
Plotting an `lm` object. There is a special `plot()` method for `lm` objects that produces
four diagnostic plots from the results of a linear model analysis.

Banner of agnes(x = iris[subset, 1:4])

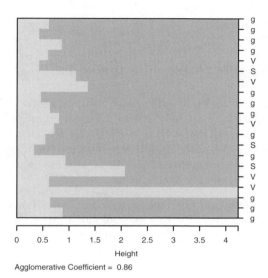

Dendrogram of agnes(x = iris[subset, 1:4])

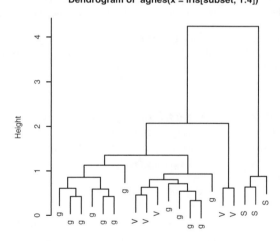

Figure 2.3
Plotting an `agnes` object. There is a special `plot()` method for `agnes` objects that produces plots relevant to the results of an agglomerative hierarchical clustering analysis.

Some add-on graphics packages provide useful extensions on the standard plot types. For example, the `Hmisc` package[26] provides the `labcurve()` function for drawing a plot with lines through multiple data series and text labels attached to each line.

2.2.1   Arguments to graphics functions

It is often the case, especially when producing graphics for publication, that the output produced by a single call to a high-level graphics function is not exactly right. There are many ways in which the output of graphics functions may be modified and Chapter 3 addresses this topic in full detail. This section will only consider the possibility of specifying arguments to high-level graphics functions in order to modify their output.

Many of these arguments are specific to a particular function. For example, the `boxplot()` function has `width` and `boxwex` arguments (among others) for controlling the width of the boxes in the plot, and the `barplot()` function has a `horiz` argument for controlling whether bars are drawn horizontally rather than vertically.

The following code shows examples of the use of the `boxwex` argument for `boxplot()` and the `horiz` argument for `barplot()` (see Figure 2.4).*

In the first example, there are two calls to `boxplot()`, which are identical except that the second specifies that the individual boxplots should be half as wide as they would be by default (`boxwex=0.5`).

```
> boxplot(decrease ~ treatment, data = OrchardSprays,
          log = "y", col="light grey")
> boxplot(decrease ~ treatment, data = OrchardSprays,
          log = "y", col="light grey",
          boxwex=0.5)
```

In the second example, there are two calls to `barplot()`, which are identical except that the second specifies that the bars should be drawn horizontally rather than vertically (`horiz=TRUE`).

---

*The data in the boxplot example are from an experiment to test the effectiveness of different orchard spray constituents in repelling honeybees, available as the data set `OrchardSprays` in the `datasets` package. The data used in the barplot example are from the `VADeaths` data set (see page 3).

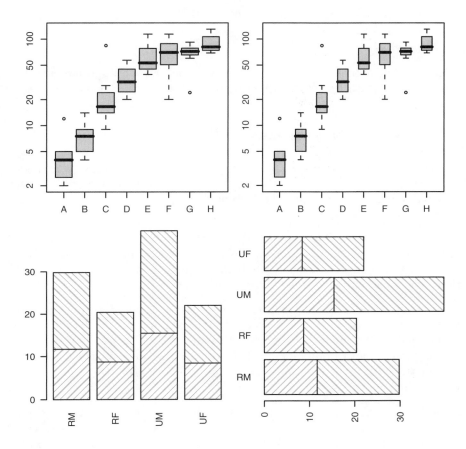

Figure 2.4
Modifying default `barplot()` and `boxplot()` output. The top two plots are produced
by calls to the `boxplot()` function with the same data, but with different values of
the `boxwex` argument. The bottom two plots are both produced by calls to the
`barplot()` function with the same data, but with different values of the `horiz`
argument.

```
> barplot(VADeaths[1:2,], angle = c(45, 135),
          density = 20, col = "grey",
          names=c("RM", "RF", "UM", "UF"))
> barplot(VADeaths[1:2,], angle = c(45, 135),
          density = 20, col = "grey",
          names=c("RM", "RF", "UM", "UF"),
          horiz=TRUE)
```

In general, the user should consult the documentation for the specific function to determine which arguments are available and what effect they have.

2.2.2   Standard arguments

Despite the existence of many arguments that are specific only to a single graphics function, there are several arguments that are "standard" in the sense that many high-level functions will accept them.

Most high-level functions will accept graphical parameters controlling such things as color (col), line type (lty), and text font. Section 3.2 provides a full list of these arguments and describes their effects. In many cases, these arguments are not given as explicitly named arguments to the high-level function, but are accepted via the ellipsis argument (. . .).

Unfortunately, because the interpretation of these standard arguments may vary in some cases, some care is necessary. For example, if the col argument is specified for a standard scatterplot, this only affects the color of the data symbols in the plot (it does not affect the color of the axes, or the axis labels), but for the barplot() function, col specifies the color for the fill or pattern used within the bars.

In addition to the standard graphical parameters, there are standard arguments to control the appearance of axes and labels on plots. It is usually possible to modify the range of the axis scales on a plot by specifying xlim or ylim arguments in the call to the high-level function, and often there is a set of arguments for specifying the labels on a plot: main for a title, sub for a sub-title, xlab for an x-axis label and ylab for a y-axis label.

Although there is no guarantee that these standard arguments will be accepted by high-level functions in add-on graphics packages, in many cases they will be accepted, and they will have the expected effect.

The following code shows examples of setting some of these standard arguments for the plot() function (see Figure 2.5). All of the calls to plot() draw a scatterplot with lines connecting the data values: the first call uses a wider line (lwd=3), the second call draws the line a grey color (col="grey"), the third call draws a dashed line (`lty="dashed"`), and the fourth call uses a much wider range of values on the y-scale (`ylim=c(-4, 4)`).

```
> y <- rnorm(20)
> plot(y, type="l", lwd=3)
> plot(y, type="l", col="grey")
> plot(y, type="l", lty="dashed")
> plot(y, type="l", ylim=c(-4, 4))
```

In cases where the default output from a high-level function cannot be modified to produce the desired result by specifying arguments to the high-level function, possible options are to add further annotation (see Section 3.4), or to generate the entire plot from scratch (see Section 3.5).

Some high-level functions provide an argument to inhibit some of the default output in order to assist in the customization of a plot. For example, the default `plot()` function has an `axes` argument to allow the user to inhibit the drawing of axes and the user can then produce customized output to represent the axis (see Section 3.4.5).

2.3 Plots of multiple variables

The traditional graphics system provides a number of functions for visualizing high-dimensional data. For plots of three variables there are: the `persp()` function for producing 3D surfaces; `contour()` and `filled.contour()` for producing contours to represent the values of the third variable; `image()`, which produces a grid of rectangles and uses color to represent the value of the third variable; and `symbols()`, which uses a symbol (e.g., a circle of varying radius) to represent the third variable. Figure 2.6 shows some examples of the output from these functions.*

For the special case of two dichotomous variables grouped by a third variable (data from a 2 by 2 by k contingency table), there is the `fourfoldplot()` function, which creates a "fourfold display"[23].

---

*The data used to produce the 3-D surface, contour, and image plots are topographic measurements of Maunga Whau (Mt. Eden), a dormant volcano in Auckland, New Zealand, available as the data set `volcano` in the **datasets** package. The data were digitally captured from a topographic map by Ross Ihaka. The data used for the `symbols()` plot are physical measurements of black cherry trees, available as the `trees` data set in the **datasets** package.

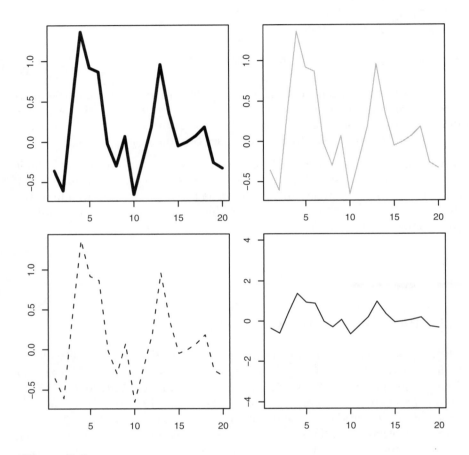

Figure 2.5

Standard arguments for high-level functions. All four plots are produced by calls to
the `plot()` function with the same data, but with different standard plot function
arguments specified: the top-left plot makes use of the `lwd` argument to control line
thickness; the top-right plot uses the `col` argument to control line color; the bottom-
left plot makes use of the `lty` argument to control line type; and the bottom-right
plot uses the `ylim` argument to control the scale on the y-axis.

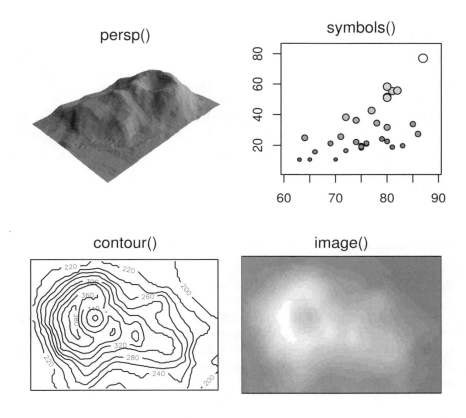

Figure 2.6
Plotting three variables. Clockwise from top-left: a 3D surface, a plot where a third variable is represented by the size of the plotting symbols, an image plot where a third variable is represented using color, and a contour plot.

For data sets containing more than three variables, there is the `pairs()` function for producing a matrix of scatterplots (plotting each variable against all other variables), the function `stars()` for producing "star" plots of continuous variables, and the `mosaicplot()` function for producing a mosaicplot of categorical data[28][24]. Figure 2.7 shows examples of the output of these functions.*

Some important add-on graphics packages provide more extensive facilities for producing representations of multi-dimensional data. For 3D plots, there is the `scatterplot3d` package[35] and the `rgl` package[2]. The latter provides some access to the visualization capabilities of OpenGL so there are advanced visualization features like the ability to interactively rotate plots and special lighting and surface effects. The `Rggobi` package[33] provides an interface between R and the `ggobi` program[57], which offers a number of techniques for visualizing many variables, including the grand tour[17].

The standard arguments described in the previous section for standard plots of one or two variables are less well supported for plotting three or more variables.

2.4 Modern plots and specialized plots

The traditional graphics system, and add-on graphics packages that have built on it, contain a number of functions to produce plots that are relatively modern (i.e., not provided by all statistical software packages), or that are suited to a particular type of data or analysis technique, or that are specific to a particular area of research.

The traditional system has functions that implement several of the plots developed by Bill Cleveland based on principles of human perception. The `dotchart()` function creates a dotplot (see the top-left plot in Figure 2.8[†]) and the `coplot()` function creates a conditioning plot (an example is shown in Figure 3.28). For a much wider range of plots of this kind, see Chapter 4, which describes Trellis plots.

---

*The data used for the scatterplot matrix are the iris data (see page 29). The data used in the `stars()` plot are measures of fuel consumption and automobile design, available as the `mtcars` data set in the `datasets` package. The data used for the mosaicplot are records of survival rates and demographic measures for passengers on the Titanic, available as the `Titanic` data set in the `datasets` package.

[†]The data set used in this example are from the `VADeaths` data set (see page 3).

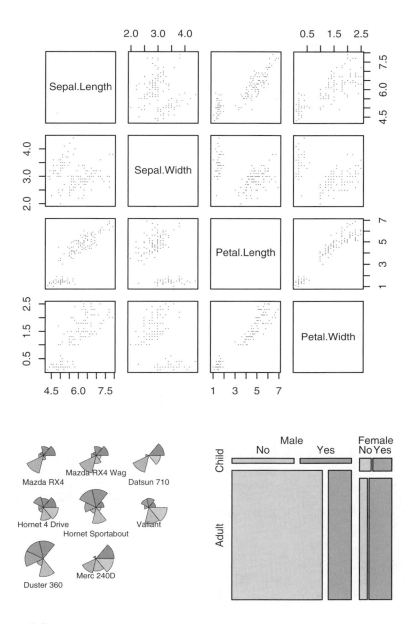

Figure 2.7
Plotting multivariate data. At the top is a scatterplot matrix (a scatterplot for every combination of a set of variables), at bottom-left is a variation on a star plot, and at bottom-right is a mosaicplot.

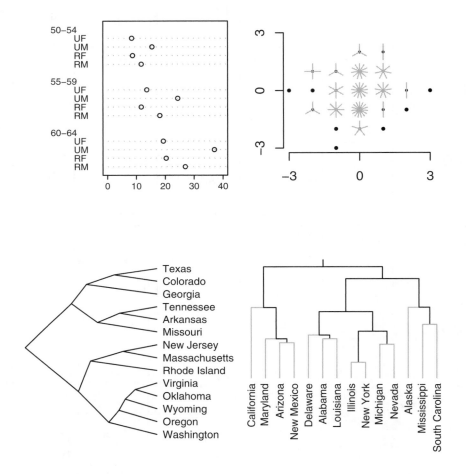

Figure 2.8
Some modern and specialized plots. Clockwise from top-left are: a (Cleveland)
dotplot, a sunflower plot, and two variations on a dendrogram.

There are several functions for helping to plot data when data symbols overlap in a standard scatterplot. The `sunflowerplot()` function can be useful when identical data values repeat a small number of times. In these plots, a "flower" is produced at each location with a "petal" for each replication (see the top-right plot in Figure 2.8). When the data are voluminous, the `hexbin()` function from the `hexbin` package is useful for plotting regions representing data density rather than plotting individual data points. A similar approach is to bin the data and use one of `contour()` or `image()`. It is also worth mentioning the `jitter()` function, which does no drawing, but adds a very small random amount to data values in order to separate values that are originally identical.

There are functions that are particularly aimed at representing categorical data or the results of analyzing categorical data. For example, the `assocplot()` function produces Cohen-Friendly association plots[16][22]. A much wider range of such functions is provided by the `vcd` package, which implements plots from Michael Friendly's book "Visualizing Categorical Data"[25].

The `plot()` method for `dendrogram` objects is provided for drawing hierarchical or tree-like structures, such as the results from clustering or a recursive partitioning regression tree. The packages `rpart`[59] and `maptree`[66] provide more functions related to this area. The bottom two plots in Figure 2.8 show examples of output from the `plot()` method for `dendrogram` objects.*

2.5 Interactive graphics

The strength of the traditional graphics system lies in the production of static graphics. There are only limited facilities for interacting with graphical output.

The `locator()` function allows the user to click within a plot and returns the coordinates where the mouse click occurred. It will also optionally draw data symbols at the clicked locations or draw lines between the clicked locations.

The `identify()` function can be used to add labels to data symbols on a plot. The data point closest to the mouse click gets labelled.

In R version 2.1.0, there is a `getGraphicsEvent()` function that provides

---

*The data used in these examples are measures of crime rates in various US states in 1973, available as the data set `USArrests` in the `datasets` package.

a more flexible basis for developing interactive plots (currently only for the Windows graphics device). This function captures key stroke events as well as mouse events and allows more general event handlers to be written as R functions.

Several add-on graphics packages provide additional interactive capabilities. The tcltk package provides a general facility for building GUI components and this can be used to create interactive graphics. Some of the tcltk demos and the dynamicGraph package[4] provide examples of this approach. The Rggobi package[33] and the iPlots package[62] provide an alternative approach by linking R to other graphics software applications that have sophisticated interactive features, such as brushing and linking plots[14][58].

Chapter summary

The traditional graphics system has functions to produce the standard statistical plots such as histograms, scatterplots, barplots, and piecharts. There are also functions for producing higher-dimensional plots such as 3D surfaces and contour plots and more specialized or modern plots such as dotplots, dendrograms, and mosaicplots. In most cases, the functions provide a number of arguments to allow the user to control the details of the plot, such as the widths of the boxes in a boxplot. There are a standard set of arguments for controlling the appearance of the plot (colors, fonts, line types, etc.) and the labels and axes on a plot, but these are not all available for all types of plots.

3

Customizing Traditional Graphics

Chapter preview

It is very often the case that a high-level plotting function does not produce exactly the final result that is desired. This chapter describes *low-level* traditional functions that are useful for controlling the fine details of a plot and for adding further output to a plot (e.g., adding descriptive labels).

In order to utilise these low-level functions effectively, this chapter also includes a description of the regions and coordinate systems that are used to locate the output from low-level functions. For example, there is a description of which function to use to draw text in the margins of a plot as opposed to drawing text in the data region (where the data symbols are plotted). There is also a discussion of ways to arrange several plots together on a single page.

Sometimes it is not possible to achieve a final result by modifying an existing high-level plot. In such cases, the user might need to create a plot using only low-level functions. This case is also addressed in this chapter together with some discussion of how to write a new graphics function for other people to use.

It is often the case that the default or standard output from a high-level function is not exactly what the user requires, particularly when producing graphics for publication. Various aspects of the output often need to be modified or completely replaced. This chapter describes the various ways in which the output from a traditional graphics high-level function can be customized and extended.

The real power of the traditional graphics system lies in the ability to control many aspects of the appearance of a plot, to add extra output to a plot, and even to build a plot from scratch in order to produce precisely the right final output.

Section 3.1 introduces important concepts of drawing regions, coordinate systems, and graphics state that are required for properly working with traditional graphics at a lower level. Section 3.2 describes how to control aspects of output such as colors, fonts, line styles, and plotting symbols, and Section 3.3 addresses the problem of placing several plots on the same page. Section 3.4 describes how to customize a plot by adding extra output and Section 3.5 looks at ways to produce entirely new types of plots.

3.1   The traditional graphics model in more detail

In order to explain some of the facilities for customizing plots, it is necessary to describe more about the model underlying traditional graphics plots.

3.1.1   Plotting regions

In the base graphics system, every page is split up into three main regions: the *outer margins*, the current *figure region*, and the current *plot region*. Figure 3.1 shows these regions when there is only one figure on the page and Figure 3.2 shows the regions when there are multiple figures on the page.

The region obtained by removing the outer margins from the device is called the *inner region*. When there is only one figure, this usually corresponds to the figure region, but when there are multiple figures the inner region corresponds to the union of all figure regions.

The area outside the plot region, but inside the figure region is referred to as the *figure margins*. A typical high-level function draws data symbols and lines within the plot region and axes and labels in the figure margins or outer margins (see Section 3.4 for information on the functions used to draw output in the different regions).

The size and location of the different regions is controlled either via the `par()` function, or using special functions for arranging plots (see Section 3.3). Specifying an arrangement of the regions does not usually affect the current plot as the settings only come into effect when the next plot is started.

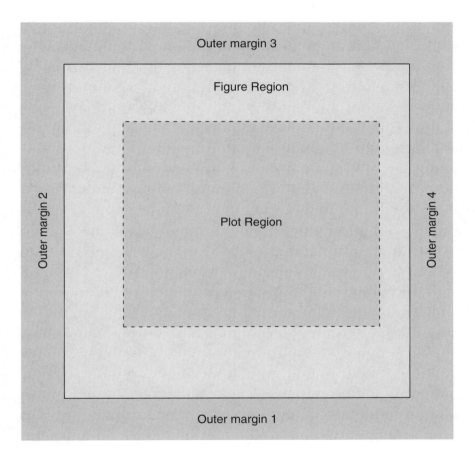

Figure 3.1
The plot regions in traditional graphics. The outer margins, figure region, and plot region, when there is a single plot on the page.

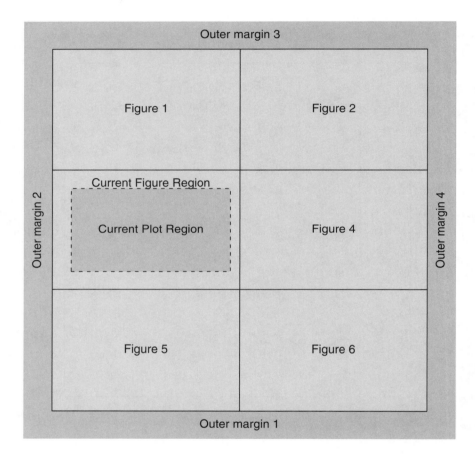

Figure 3.2

Multiple figure regions in traditional graphics. The outer margins, *current* figure region, and *current* plot region, when there are multiple plots on the page.

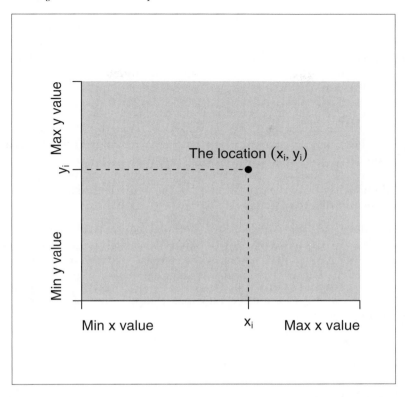

Figure 3.3
The user coordinate system in the plot region. Locations within this coordinate system are relative to the scales on the plot axes.

Coordinate systems

Each plotting region has one or more coordinate systems associated with it. Drawing in a region occurs relative to the relevant coordinate system. The coordinate system in the plot region, referred to as "user coordinates," is probably the easiest to understand as it simply corresponds to the range of values on the axes of the plot (see Figure 3.3). The drawing of data symbols, lines, text, and so on in the plot region is relative to this user coordinate system.

The scales on the axes of a plot are often set up automatically by R, but it is possible to control them explicitly via `xlim` and `ylim` arguments to high-level plotting functions (see Section 2.2.1) or via the `usr` argument to the `par()` function (see Section 3.4.7).

The figure margins contain the next most commonly-used coordinate systems. The coordinate systems in these margins are a combination of x- or y-ranges (like user coordinates) and lines of text away from the boundary of the plot region. Figure 3.4 shows two of the four figure margin coordinate systems. Axes are drawn in the figure margins using these coordinate systems.

There is a further set of "normalized" coordinate systems available for the figure margins in which the x- and y-ranges are replaced with a range from 0 to 1. In other words, it is possible to specify locations along the axes as a proportion of the total axis length. Axis labels and plot titles are drawn relative to this coordinate system. All of these figure margin coordinate systems are created implicitly from the arrangement of the figure margins and the setting of the user coordinate system.

The outer margins have similar sets of coordinate systems, but locations along the boundary of the inner region can only be specified in normalized coordinates (always relative to the extent of the complete outer margin). Figure 3.5 shows two of the four outer margin coordinate systems.

Sections 3.4.3 and 3.4.5 describe functions that produce output relative to these margin coordinate systems.

3.1.2 The traditional graphics state

The traditional graphics system maintains a graphics "state" for each graphics device. Whenever output is drawn, the graphics state is consulted to determine where it should be drawn, what color it should be, what fonts to use for text, and so on.

The graphics state consists of a large number of settings. Some of these settings describe the size and placement of the plot regions and coordinate systems described above. Some settings describe the general appearance of graphical output (the colors and line types that are used to draw lines, the fonts that are used to draw text, etc). Some settings describe aspects of the output device (e.g., the physical size of the device and the current clipping region).

Tables 3.1 to 3.3 together provide a list of all of the graphics state settings and a very brief indication of their meaning. Most of the settings are described in detail in Sections 3.2 and 3.3.

The main function used to access the graphics state is the **par()** function. Simply typing **par()** will result in a complete listing of the current graphics state. A specific state setting can be queried by supplying specific setting names as arguments to **par()**. The following code (page 52) queries the current state of the **col** and **lty** settings.

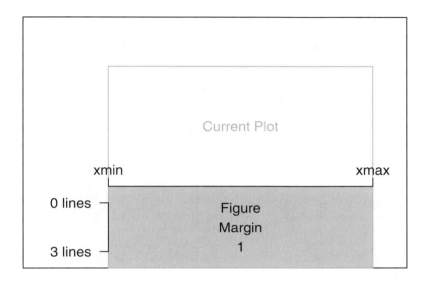

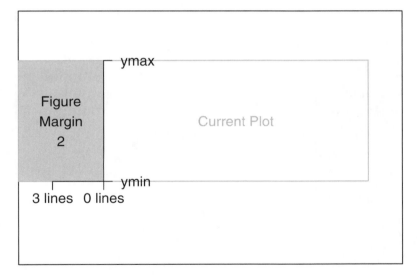

Figure 3.4
Figure margin coordinate systems. The typical coordinate systems for figure margin 1 (top plot) and figure margin 2 (bottom plot). Locations within these coordinate systems are a combination of position along the axis scale and distance away from the axis in multiples of lines of text.

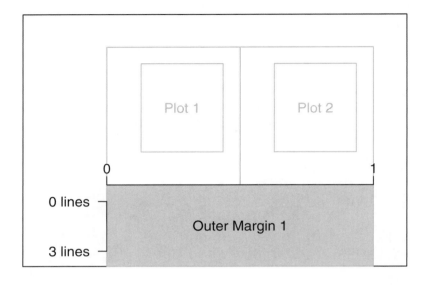

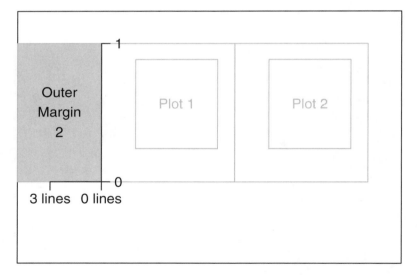

Figure 3.5
Outer margin coordinate systems. The typical coordinate systems for outer margin
1 (top plot) and outer margin 2 (bottom plot). Locations within these coordinate
systems are a combination of a proportion along the inner region and distance away
from the inner region in multiples of lines of text.

Table 3.1

High-level traditional graphics state settings. This set of graphics state settings can be queried and set via the `par()` function *and* can be used as arguments to other graphics functions (e.g., `plot()` or `lines()`). Each setting is described in more detail in the relevant **Section**.

Setting	Description	Section
adj	justification of text	3.2.3
ann	draw plot labels and titles?	3.2.3
bg	"background" color	3.2.1
bty	type of box drawn by `box()`	3.2.5
cex	size of text (multiplier)	3.2.3
cex.axis	size of axis tick labels	3.2.3
cex.lab	size of axis labels	3.2.3
cex.main	size of plot title	3.2.3
cex.sub	size of plot sub-title	3.2.3
col	color of lines and data symbols	3.2.1
col.axis	color of axis tick labels	3.2.1
col.lab	color of axis labels	3.2.1
col.main	color of plot title	3.2.1
col.sub	color of plot sub-title	3.2.1
fg	"foreground" color	3.2.1
font	font face (bold, italic) for text	3.2.3
font.axis	font face for axis tick labels	3.2.3
font.lab	font face for axis labels	3.2.3
font.main	font face for plot title	3.2.3
font.sub	font face for plot sub-title	3.2.3
gamma	gamma correction for colors	3.2.1
lab	number of ticks on axes	3.2.5
las	rotation of text in margins	3.2.3
lty	line type (solid, dashed)	3.2.2
lwd	line width	3.2.2
mgp	placement of axis ticks and tick labels	3.2.5
pch	data symbol type	3.2.4
srt	rotation of text in plot region	3.2.3
tck	length of axis ticks (relative to plot size)	3.2.5
tcl	length of axis ticks (relative to text size)	3.2.5
tmag	size of plot title (relative to other labels)	3.2.3
type	type of plot (points, lines, both)	3.2.4
xaxp	number of ticks on x-axis	3.2.5
xaxs	calculation of scale range on x-axis	3.2.5
xaxt	x-axis style (standard, none)	3.2.5
xpd	clipping region	3.2.7
yaxp	number of ticks on y-axis	3.2.5
yaxs	calculation of scale range on y-axis	3.2.5
yaxt	y-axis style (standard, none)	3.2.5

```
> par(c("col", "lty"))

$col
[1] "black"

$lty
[1] "solid"
```

The `par()` function can be used to modify traditional graphics state settings by specifying a value via an argument with the appropriate setting name. The following code sets new values for the `col` and `lty` settings.

```
> par(col="red", lty="dashed")
```

Modifying traditional graphics state settings via `par()` has a persistent effect. Settings specified in this way will hold until a different setting is specified. Settings may also be *temporarily* modified by specifying a new value in a call to a graphics function such as `plot()` or `lines()`. The following code demonstrates this idea. First of all, the line type is permanently set using `par()`, then a plot is drawn and the lines drawn between data points in this plot are dashed. Next, a plot is drawn with a temporary line type setting of `lty="solid"` and the lines in this plot are solid. When the third plot is drawn, the permanent line type setting of `lty="dashed"` is back in effect so the lines are again dashed.

```
> y <- rnorm(20)
> par(lty="dashed")
> plot(y, type="l") # line is dashed
> plot(y, type="l", lty="solid") # line is solid
> plot(y, type="l") # line is dashed
```

Only some of the graphics state settings can be set temporarily in calls to graphics functions. For example, the `mfrow` setting may not be set in this way and can only be set using `par()`. These "low level" settings are listed in Table 3.2.

A small set of graphics state settings cannot be set at all and can only be queried using `par()`. For example, there is no function to allow the user to modify the size of the current device (after the device has been created), but its size (in inches) may be obtained using `par("din")`. These "read only" settings are listed in Table 3.3.

Changes to the traditional graphics state only affect the current graphics device.

Table 3.2

Low-level traditional graphics state settings. This set of graphics state settings can be queried and set via the `par()` function. Each setting is described in more detail in the relevant **Section**.

Setting	Description	Section
ask	prompt user before new page?	3.2.8
family	font family for text	3.2.3
fig	location of figure region (normalized)	3.2.6
fin	size of figure region (inches)	3.2.6
lend	line end style	3.2.2
lheight	line spacing (multiplier)	3.2.3
ljoin	line join style	3.2.2
lmitre	line mitre limit	3.2.2
mai	size of figure margins (inches)	3.2.6
mar	size of figure margins (lines of text)	3.2.6
mex	line spacing in margins	3.2.6
mfcol	number of figures on a page	3.3.1
mfg	which figure is used next	3.3.1
mfrow	number of figures on a page	3.3.1
new	has a new plot been started?	3.2.8
oma	size of outer margins (lines of text)	3.2.6
omd	location of inner region (normalized)	3.2.6
omi	size of outer margins (inches)	3.2.6
pin	size of plot region (inches)	3.2.6
plt	location of plot region (normalized)	3.2.6
ps	size of text (points)	3.2.3
pty	aspect ratio of plot region	3.2.6
usr	range of scales on axes	3.4.7
xlog	logarithmic scale on x-axis?	3.2.5
ylog	logarithmic scale on y-axis?	3.2.5

Table 3.3

Read-only traditional graphics state settings. This set of graphics state settings can only be queried (via the `par()` function). Each setting is described in more detail in the relevant **Section**.

Setting	Description	Section
cin	size of a character (inches)	3.4.7
cra	size of a character ("pixels")	3.4.7
cxy	size of a character (user coordinates)	3.4.7
din	size of graphics device (inches)	3.4.7

3.2   Controlling the appearance of plots

This section is concerned with the appearance of plots, which means the colors, line types, fonts and so on that are used to draw a plot. As described in Section 3.1.2, these features are controlled via traditional graphics state settings and values are specified for the settings either with a call to the `par()` function or as arguments to a specific graphics function such as `plot()`. For example, there is a setting called `col` to control the color of output (see the next section). This can be set permanently using `par()` with an expression of the form

```
par(col="red")
```

which will affect all subsequent graphical output. Alternatively, the setting can be specified as an argument to a high-level function using an expression like

```
plot(..., col="red")
```

which means that the setting will affect the output just for that plot. Finally, the setting can be used as an argument to a low-level function, as in the expression

```
lines(..., col="red")
```

which shows that the setting can be used to control the appearance of a single piece of graphical output.

There are many individual settings that affect the appearance of a plot, but they can be grouped in terms of what aspects of a plot the settings affect. Each of the following sections details a particular group of settings, including a description of the role of individual settings and descriptions of what constitutes valid values for each setting. There are sections on: specifying colors; how to control the appearance of lines, text, data symbols, and axes; how to control the size and location of the various plotting regions; clipping (only drawing output on certain parts of the page); and specifying what should happen when a high-level function is called to start a new plot.

The appearance of plots is also affected by the location and size of the plotting regions, but this is dealt with separately in Section 3.3.

This section is not meant to be read from start to end as it is very detailed. This section should be used as a reference tool to access the relevant subsections as they are required to learn about controlling a particular aspect of a plot.

3.2.1 Colors

There are three main color settings in the traditional graphics state: `col`, `fg`, and `bg`.

The `col` setting is the most commonly used. The primary use is to specify the color of data symbols, lines, text, and so on that are drawn in the plot region. Unfortunately, when specified via a graphics function, the effect can vary. For example, a standard scatterplot produced by the `plot()` function will use `col` for coloring data symbols and lines, but the `barplot()` function will use `col` for filling the contents of its bars. In the `rect()` function, the `col` argument provides the color to fill the rectangle and there is a `border` argument specific to `rect()` that gives the color to draw the border of the rectangle. The effect of `col` on graphical output drawn in the margins also varies. It does not affect the color of axes and axis labels, but it does affect the output from the `mtext()` function. There are specific settings for affecting axes, labels, titles, and sub-titles called `col.axis`, `col.lab`, `col.main`, and `col.sub`.

The `fg` setting is primarily intended for specifying the color of axes and borders on plots. There is some overlap between this and the specific `col.axis`, `col.main`, etc. settings described above.

The `bg` setting is primarily intended to specify the color of the background for base graphics output. This color is used to fill the entire page. As with the `col` setting, when `bg` is specified in a graphics function it can have a quite different meaning. For example, the `plot()` and `points()` function use `bg` to specify the color for the interior of the data symbols, which can have different colors on the border (`pch` values 21 to 25; see Section 3.2.4).

There is also a `gamma` setting that controls the gamma correction for a device. On most devices this can only be set once when the device is first opened.

Specifying colors

The easiest way to specify a color in R is simply to use the color's name. For example, `"red"` can be used to specify that graphical output should be (a very bright) red. R understands a fairly large set of color names (657 to be exact); type `colors()` (or `colours()`) to see a full list of known names.

It is also possible to specify colors using one of the standard color-space descriptions. For example, the `rgb()` function allows a color to be specified as a Red-Green-Blue (RGB) triplet of intensities. Using this function, the color red is specified as `rgb(1, 0, 0)` (i.e., as much red as possible, no blue, and no green). The function `col2rgb()` can be used to see the RGB values for a particular color name.

An alternative way to provide an RGB color specification is to provide a string of the form `"#RRGGBB"`, where each of the pairs `RR`, `GG`, `BB` consist of two hexadecimal digits giving a value in the range zero (`00`) to 255 (`FF`). In this specification, the color red is given as `"#FF0000"`.

There is also an `hsv()` function for specifying a color as a Hue-Saturation-Value (HSV) triplet. The terminology of color spaces is fraught, but roughly speaking: hue corresponds to a position on the rainbow, from red (0), through orange, yellow, green, blue, indigo, to violet (1); saturation determines whether the color is dull or bright; and value determines whether the color is light or dark. The HSV specification for the (very bright) color red is `hsv(0, 1, 1)`. The function `rgb2hsv()` converts a color specification from RGB to HSV.

There is also a `convertColor()` function for converting colors between different color spaces, including the CIELAB and CIELUV color spaces[46], in which a unit distance represents a perceptually constant change in color. The `hcl()` function allows colors to be specified directly as polar coordinates within CIELUV (as a hue, chroma, and luminance triplet). This is like a perceptually uniform version of HSV.* Ross Ihaka's `colorspace` package[31] provides an alternative set of functions for generating, converting, and combining colors in a sophisticated manner in a wide variety of color spaces.

One final way to specify a color is simply as an integer index into a predefined set of colors. The predefined set of colors can be viewed and modified using the `palette()` function. In the default palette, red is specified as the integer 2.

Semitransparent colors

All R colors are stored with an alpha transparency channel. An alpha value of 0 means fully transparent and an alpha value of 1[†] means fully opaque. When an alpha value is not specified, the color is opaque.

The function `rgb()` can be used to specify a color with an alpha transparency

---

*The `hcl()` function is only available from R version 2.1.0.

†The maximum alpha value depends on the method being used to specify a color. When a color is specified via `rgb()`, the user can decide what the maximal value should be (it defaults to 1). When a color is specified as a string beginning with a `"#"`, the maximum value is `"FF"`.

Table 3.4

Functions to generate color sets. R functions that can be used to generate coherent sets of colors

Name	Description
rainbow()	Colors vary from red through orange, yellow, green, blue, and indigo, to violet.
heat.colors()	Colors vary from white, through orange, to red.
terrain.colors()	Colors vary from white, through brown, to green.
topo.colors()	Colors vary from white, through brown then green, to blue.
cm.colors()	Colors vary from light blue, through white, to light magenta.
grey() or gray()	A set of shades of grey.

channel (e.g., `rgb(1, 0, 0, 0.5)` specifies a semitransparent red), or a color can be specified as a string beginning with a `"#"` and followed by eight hexadecimal digits. In the latter case, the last two hexadecimal digits specify an alpha value in the range 0 to 255 (e.g., `"#FF000080"` specifies a semitransparent red).

A color may also be specified as `NA`, which is usually interpreted as fully transparent (i.e., nothing is drawn). The special color name `"transparent"` can also be used to specify full transparency.

Only the PDF and Quartz devices support semitransparent colors. On all other devices, semitransparent colors are rendered as fully transparent.

Color sets

More than one color is often required within a single plot and in such cases it can be difficult to select colors that are aesthetically pleasing or are related in some way (e.g., a set of colors in which the brightness of the colors decreases in regular steps). Table 3.4 lists some functions that R provides for generating sets of colors. The output of the expression `example(rainbow)` provides a nice visual summary of the color sets generated by several of these functions.

Each of the functions in Table 3.4 selects a set of colors by taking regular steps along a path through the HSV color space. This can produce color sets that do not appear to vary smoothly. A perceptually constant color space makes it easier to generate sets of colors with even perceptual steps between them or a set of colors that do not vary on a particular perceptual dimension. For example, the following code generates six colors from the CIELUV color space that vary regularly in terms of hue, but are all equally bright (the chroma component is fixed at 50) and all equally light (the luminance component is fixed at 60).

```
> hcl(seq(0, 360, length=7)[-7], 50, 60)

[1] "#C87A8A" "#AC8C4E" "#6B9D59" "#00A396" "#5F96C2"
[6] "#B37EBE"
```

The RColorBrewer package[47] provides color palettes from Cynthia Brewer's ColorBrewer tool[27]. The ColorBrewer color sets have been carefully selected by a color expert and include distinct palettes for representing nominal and ordinal categories.

The functions colorRamp() and colorRampPalette() can be used to inter-polate a new color set from an existing set of colors (e.g., create additional colors from within a ColorBrewer palette).*

Device Dependency of Color Specifications

R stores colors internally as RGB triplets. The final appearance of a color can vary considerably when it is viewed on a screen, or printed on paper, or displayed through a projector as it depends on the physical characteristics of the screen, printer ink, or projector.

Fill patterns

In some cases (e.g., when printing in black and white), it is difficult to make use of different colors to distinguish between different elements of a plot. Using different levels of grey can be effective, but another option is to make use of some sort of fill pattern, such as cross-hatching. These should be used with caution because it is very easy to create visual effects that are distracting. Nevertheless, some journals actively encourage their use, so the facility has some purpose.

In R, there is only limited support for fill patterns and they can only be applied to rectangles and polygons (and only within the traditional graphics

---

*The functions colorRamp(), colorRampPalette(), and convertColor() are not avail-able before R version 2.1.0, but some color ramp functionality is available in the hexbin package[10], which is part of the BioConductor project.

system). It is possible to fill a rectangle or polygon with a set of lines drawn at a certain angle, with a specific separation between the lines. A `density` argument controls the separation between the lines (in terms of lines per inch) and an `angle` argument controls the angle of the lines (in terms of degrees anti-clockwise from 3 o'clock). Examples of the use of fill patterns are given in Figures 2.4, 3.20, and their associated code.

These settings can only be controlled via arguments to the functions `rect()`, `polygon()`, `hist()`, `barplot()`, `pie()`, and `legend()` (and *not* via `par()`).

3.2.2 Lines

There are five graphics state settings for controlling the appearance of lines. The `lty` setting describes the type of line to draw (solid, dashed, dotted, ...), the `lwd` setting describes the width of lines, and the `ljoin`, `lend`, and `lmitre` settings control how the ends and corners in lines are drawn (see below).

The scope of these settings again differs depending on the graphics function being called. For example, for standard scatterplots, the setting only applies to lines drawn within the plot region. In order to affect the lines drawn as part of the axes, the `lty` setting must be passed directly to the `axis()` function.

Specifying line widths

The width of lines is specified by a simple numeric value, e.g., `lwd=3`. The interpretation of this value depends on what sort of device the line is being drawn on. In other words, the physical width of the line may be different when the line is drawn on a computer screen compared to when it is printed on a sheet of paper. On a computer screen, a line width of 1 will typically mean one pixel. For PostScript and PDF output, a line width of 1 produces a line 0.75 points wide. The default value is 1.

Specifying line types

R graphics supports a fixed set of predefined line types, which can be specified by name, such as `"solid"` or `"dashed"`, or as an integer index (see Figure 3.6). In addition, it is possible to specify customized line types via a string of digits. In this case, each digit is a hexadecimal value that indicates a number of "units" to draw either a line or a gap. Odd digits specify line lengths and even digits specify gap lengths. For example, a dotted line is specified by `lty="13"`, which means draw a line of length one unit then a gap of length three units. A unit corresponds to the current line width, so the result scales with line width, but is device-dependent. Up to four such line-gap pairs can be specified. Figure 3.6 shows the available predefined line types and some examples of customized line types.

Specifying line ends and joins

When drawing thick lines, it becomes important to select the style that is used to draw corners (joins) in the line and the ends of the line. R provides three styles for both cases: there is an `lend` setting to control line ends, which can be `"round"` or flat (with two variations on flat, `"square"` or `"butt"`); and there is an `ljoin` setting to control line joins, which can be `"mitre"` (pointy), `"round"`, or `"bevel"`. The differences are most easily demonstrated visually (see Figure 3.7).

When the line join style is `"mitre"`, the join style will automatically be converted to `"bevel"` if the angle at the join is too small. This is to avoid excessively pointy joins. The point at which the automatic conversion occurs is controlled by a setting called `lmitre`, which specifies the ratio of the length of the mitre divided by the line width. The default value is `10`, which means that the conversion occurs for joins where the angle is less than 11 degrees. Other standard values are `2`, which means that conversion occurs at angles less than 60 degrees, and `1.414`, which means that conversion occurs for angles less than 90 degrees. The minimum mitre limit value is `1`.

These settings can only be specified via `par()` (not as arguments to high-level or low-level graphics functions) and not all devices will respect them (especially the line mitre limit).

It is important to remember that line join styles influence the corners on rectangles and polygons as well as joins in lines.

3.2.3   Text

There are a large number of traditional graphics state settings for controlling the appearance of text. The size of text is controlled via `ps` and `cex`; the font is controlled via `font` and `family`; the justification of text is controlled via `adj`; and the angle of rotation is controlled via `srt`.

There is also an `ann` setting, which indicates whether titles and axis labels should be drawn on a plot. This is intended to apply to high-level functions, but is not guaranteed to work with all such functions (especially functions from add-on graphics packages). There are examples of the use of `ann` as an argument to high-level plotting functions in Section 3.4.

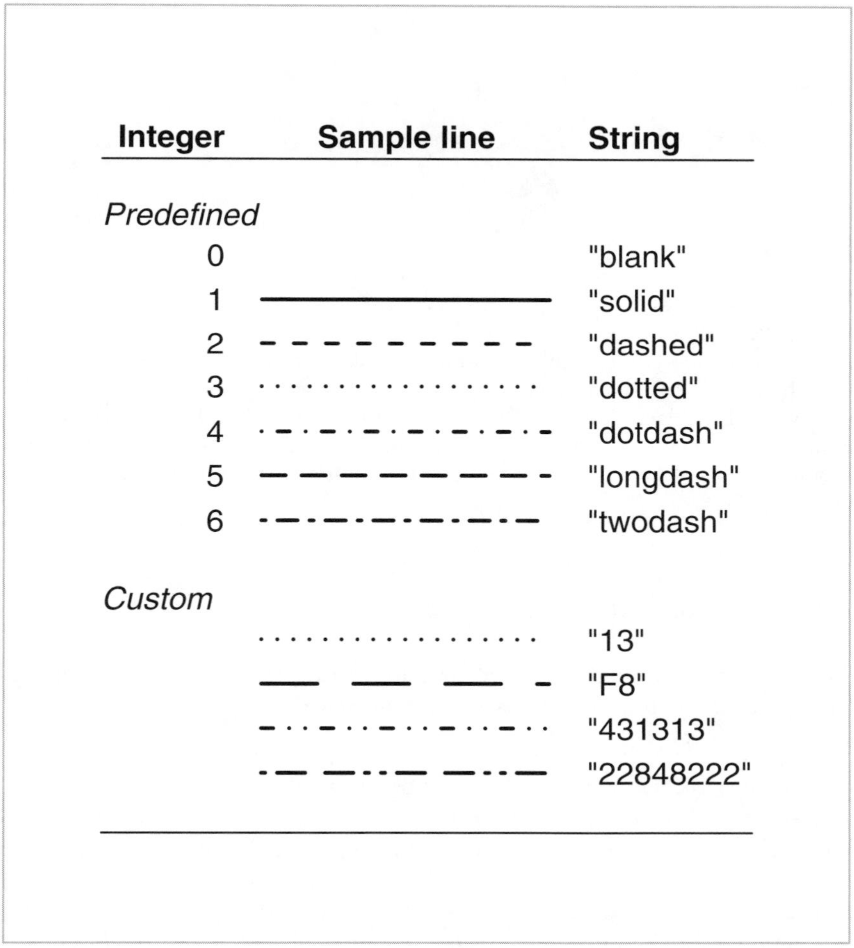

Integer	Sample line	String
Predefined		
0		"blank"
1	————————	"solid"
2	– – – – – – – – –	"dashed"
3	· · · · · · · · · · · · · · · ·	"dotted"
4	· – · – · – · – · – · · –	"dotdash"
5	— — — — — — — –	"longdash"
6	· – · – · – · – · –	"twodash"
Custom		
	· · · · · · · · · · · · · · · ·	"13"
	— — — –	"F8"
	– · · – · · – · · – · ·	"431313"
	· — —· · — —· · —	"22848222"

Figure 3.6
Predefined and custom line types. Line type may be specified as a predefined integer, as a predefined string name, or as a string of hexadecimal characters specifying a custom line type.

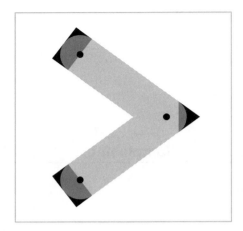

Figure 3.7
Line join and line ending styles. Three thick lines have been drawn through the
same three points (indicated by black circles), but with different line end and line
join styles. The black line was drawn first with `"square"` ends and `"mitre"` joins;
the dark grey line was drawn on top of the black line with `"round"` ends and `"round"`
joins; and the light grey line was drawn on top of that with `"butt"` ends and `"bevel"`
joins.

Justification of text

The `adj` setting is a value from 0 to 1 indicating the horizontal justification
of text strings (0 means left-justified, 1 means right-justified and a value of
0.5 centers text).

The meaning of the `adj` setting depends on whether text is being drawn in
the plot region, in the figure margins, or in the outer margins. In the plot
region, the justification is relative to the (x, y) location at which the text
is being drawn. In this context, it is also possible to specify two values for
the setting and the second value is taken as a vertical justification for the
text. Furthermore, non-finite values (`NA`, `NaN`, or `Inf`) may be specified for
the justification and this is taken to mean "exact" centering. There is only
a difference between a justification value of 0.5 and a non-finite justification
value for vertical justification. In this case, a setting of 0.5 means text is
vertically centered based on the height of the text above the text baseline
(i.e., ignoring "descenders" like the tail on a "y"). A non-finite value means
that text is vertically centered based on the full height of the text (including
descenders). Figure 3.8 shows how various `adj` settings affect the alignment
of text in the plot region.

In the figure margins and outer margins, the meaning of the `adj` setting

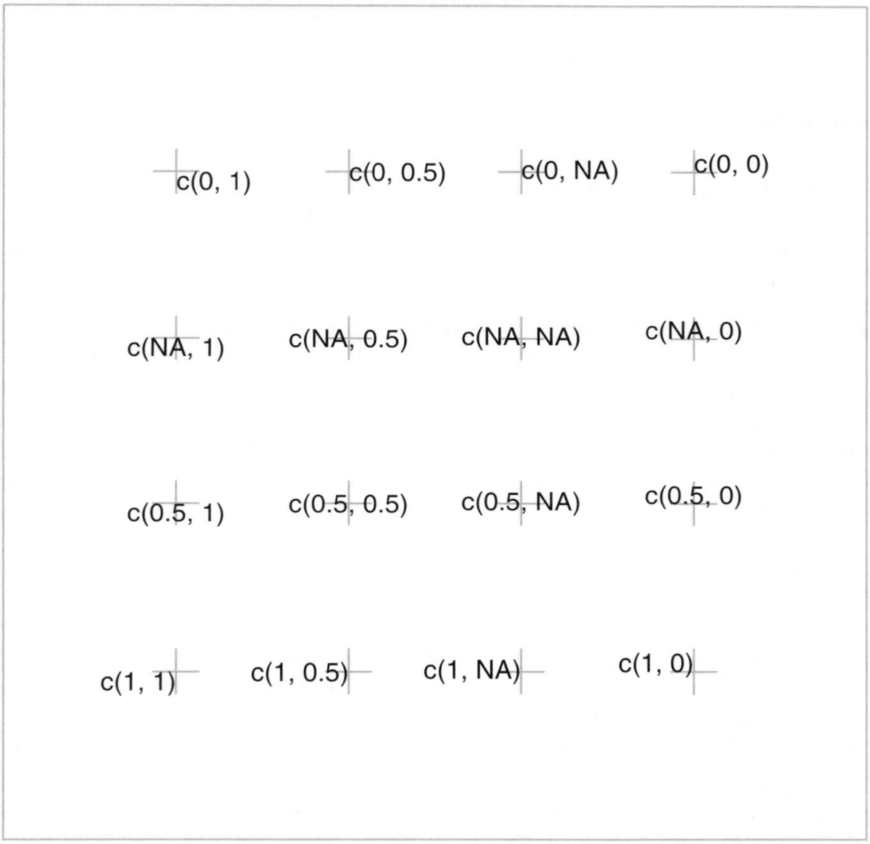

Figure 3.8
Alignment of text in the plot region. The `adj` graphical setting may be given two values, c(*hjust*, *vjust*), where *hjust* specifies horizontal justification and *vjust* specifies vertical justification. Each piece of text in the diagram is justified relative to a grey cross to represent the effect of the relevant `adj` setting. The vertical adjustment for `NA` is subtly different from the vertical adjustment for 0.5.

depends on the `las` setting. When margin text is parallel to the axis, `adj` specifies *both* the location and the justification of the text. For example, a value of 0 means that the text is left-justified *and* that the text is located at the left end of the margin. When text is perpendicular to the axis, the `adj` setting only affects justification. Furthermore, the `adj` setting only affects "horizontal" justification (justification in the reading direction) for text in the margins.

Rotating text

The `srt` setting specifies a rotation angle anti-clockwise from the positive x-axis, in degrees (not radians). This will only affect text drawn in the plot region (text drawn by the `text()` function). Text can be drawn at any angle within the plot region.

In the figure and outer margins, text may only be drawn at angles that are multiples of 90°, and this angle is controlled by the `las` setting. A value of 0 means text is always drawn parallel to the relevant axis (i.e., horizontal in margins 1 and 3, and vertical in margins 2 and 4). A value of 1 means text is always horizontal, 2 means text is always perpendicular to the relevant axis, and 3 means text is always vertical. This setting interacts with or overrides the `adj` and `srt` settings.

Text size

The size of text is ultimately a numerical value specifying the size of the font in "points." The font size is controlled by two settings: `ps` specifies an absolute font size setting (e.g., `ps=9`), and `cex` specifies a multiplicative modifier (e.g., `cex=1.5`). The final font size specification is simply `fontsize * cex`. On some devices, the font size that is specified will not be honored exactly. For example, when drawing in an X11 window with bitmap fonts, there are only a finite set of font sizes available and this set will vary depending on which fonts are installed. For the PostScript and PDF formats, font sizes should be accurate.

As with specifying color, the scope of a `cex` setting can vary depending on where it is given. When `cex` is specified via `par()`, it affects most text. However, when `cex` is specified via `plot()`, it only affects the size of data symbols. There are special settings for controlling the size of text that is drawn as axis tick labels (`cex.axis`), text that is drawn as axis labels (`cex.lab`), text in the title (`cex.main`), and text in the sub-title (`cex.sub`). Finally, there is a `tmag` setting for controlling the amount to magnify title text relative to other plot labels.

Multi-line text

It is possible to draw text that spans several lines, by inserting a new line escape sequence, `"\n"`, within a piece of text, as in the following example.

```
"first line\nsecond line"
```

The spacing between lines is controlled by the `lheight` setting, which is a multiplier applied to the natural height of a line of text. For example, `lheight=2` specifies double-spaced text. This setting can only be specified via `par()`.

Specifying fonts

Specifying an exact font may involve several pieces of information and is very device-specific. A font is usually part of a font "family" (e.g. Helvetica or `Courier`) and is a particular "face" within that family (e.g., **bold** or *italic*). It is also possible to specify things like the font format (e.g., TrueType or Computer Modern), the font encoding (e.g., ISO Latin 1), and even the font foundry or designer (e.g., Adobe or Sun Microsystems).

In R graphics, it is possible to specify the font face and a font family. On some devices, the latter can include extra details such as encoding.

The font face is specified via the `font` setting as an integer (Table 3.5 shows the possible values). As with color and text size, the `font` setting applies only to text drawn in the plot region. There are additional settings specifically for axes (`font.axis`), labels (`font.lab`), and titles (`font.main` and `font.sub`).

Every graphics device establishes a default font family, which is usually a sans serif font such as Helvetica or Arial. A new font family is specified via the `family` setting using a device-independent name. The names `"sans"`, `"serif"`, `"mono"`, and `"symbol"` are available for the most common devices[*] and provide a sans serif font, a serif font, a monospaced font, and a symbol font respectively (see Table 3.6).

Figure 3.9 demonstrates the 16 basic font family and face combinations.[†]

The device-independent font name is mapped to a device-dependent font family by individual devices. These mappings can be modified and new font names and mappings defined using functions such as `postscriptFont()` and `postscriptFonts()`.

---

[*]Windows, X11, Quartz, PDF, and PostScript.

[†]The fact that there is a font *specification* provided for all standard devices does not mean that a matching font will always be available. There can be significant differences between operating systems and locales in terms of which fonts are installed by default.

Table 3.5
Possible font face specifications in traditional graphics. The font face must be specified as an integer, usually between 1 and 4. The special value 5 indicates that a symbol font should be used. The range of valid font faces varies for different Hershey fonts, but the maximum valid value is usually 4 or less. When the font family is `"HersheySerif"`, there are a number of special font faces available.

Integer	Description
1	Roman or upright face
2	Bold face
3	Slanted or italic face
4	Bold and slanted face
5	Symbol
For the HersheySerif font family	
5	Cyrillic font
6	Slanted Cyrillic font
7	Japanese characters

Table 3.6
Device-independent and Hershey font families that are distributed with R. A font family is specified as a string

Name	Description
Device-independent fonts	
`"serif"`	Serif variable-width font
`"sans"`	Sans-serif variable-width font
`"mono"`	Mono-spaced "typewriter" font
`"symbol"`	Symbol font
Hershey fonts	
`"HersheySerif"`	Serif variable-width font
`"HersheySans"`	Sans-serif variable-width font
`"HersheyScript"`	Serif "handwriting" font
`"HersheyGothicEnglish"`	Gothic script font
`"HersheyGothicGerman"`	Gothic script font
`"HersheyGothicItalian"`	Gothic script font
`"HersheySymbol"`	Serif symbol font
`"HersheySansSymbol"`	Sans-serif symbol font

φαμιλψ=∀μονο∀   φαμιλψ=∀μονο∀   φαμιλψ=∀μονο∀   φαμιλψ=∀μονο∀
φοντ=1   φοντ=2   φοντ=3   φοντ=4

family="mono"   **family="mono"**   *family="mono"*   ***family="mono"***
font=1   **font=2**   *font=3*   ***font=4***

family="serif"   **family="serif"**   *family="serif"*   ***family="serif"***
font=1   **font=2**   *font=3*   ***font=4***

family="sans"   **family="sans"**   *family="sans"*   ***family="sans"***
font=1   **font=2**   *font=3*   ***font=4***

Figure 3.9
Font families and font faces. The appearance of the base sixteen font family and
font face combinations that are available for X11, PDF, PostScript, Windows, and
Quartz graphics devices (the output shown is for the PostScript device).

The Hershey outline fonts[1] are also distributed with R and are available for *all* output formats. The names to use with the `family` setting to obtain the different Hershey fonts are shown in Table 3.6. See the on-line help page for `Hershey` for more information on Hershey fonts.

The `family` setting can only be specified via `par()` (not as an argument to a high-level plotting function).

Locales

From R version 2.1.0, there is support for multi-byte locales, such as UTF-8 locales and East Asian locales (Chinese, Japanese, and Korean). This means that strings can be specified in R that contain characters outside of the ISO Latin 1 character set that R was restricted to prior to version 2.1.0. Such characters cannot be produced within graphical output on all devices.

As long as the appropriate fonts are available, it should be possible to produce characters outside of the ISO Latin 1 set for X11, Windows, and Quartz devices, but PostScript and PDF output can only be produced for ISO Latin 1 characters.

3.2.4 Data symbols

R provides a fixed set of 26 data symbols for plotting and the choice of data symbol is controlled by the `pch` setting. This can be an integer value to select one of the fixed set of data symbols, or a single character (see Figure 3.10). Some of the predefined data symbols (`pch` between 21 and 25) allow a fill color separate from the border color, with the `bg` setting controlling the fill color in these cases. If `pch` is a character then that letter is used as the plotting symbol. The character `"."` is treated as a special case and the device attempts to draw a very small dot (see, for example, the scatterplot matrix in Figure 2.7).

The size of the data symbols is linked to the size of text and is affected by the `cex` setting. If the data symbol is a character, the size will also be affected by the `ps` setting.

The `type` setting controls how data is represented in a plot. A value of `"p"` means that data symbols are drawn at each (`x, y`) location. The value `"l"` means that the (`x, y`) locations are connected by lines. A value of `"b"` means that both data symbols and lines are drawn. The `type` setting may also have the value `"o"`, which means that data symbols are "over-plotted" on lines (with the value `"b"`, the lines stop short of each data symbol). It is also possible to specify the value `"h"`, which means that vertical lines are drawn from the

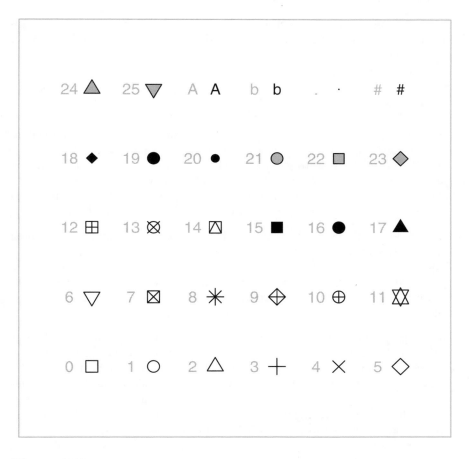

Figure 3.10
Data symbols available in R. A particular data symbol is selected by specifying an integer between 0 and 25 or a single character for the **pch** graphical setting. In the diagram, the relevant integer or character **pch** value is shown in grey to the left of the relevant symbol.

x-axis to the (x, y) locations (the appearance is like a barplot with very thin bars). Two further values, "s" and "S" mean that (x, y) locations are joined in a city-block fashion with lines going horizontally then vertically (or vertically then horizontally) between each data location. Finally, the value "n" means that nothing is drawn at all.

Figure 3.11 shows simple examples of the different plot types. This setting is most often specified within a call to a high-level function (e.g., plot()) rather than via par().

3.2.5 Axes

By default, the traditional graphics system produces axes with sensible labels and tick marks at sensible locations. If the axis does not look right, there are a number of graphical state settings specifically for controlling aspects such as the number of tick marks and the positioning of labels. These are described below. If none of these gives the desired result, the user may have to resort to drawing the axis explicitly using the axis() function (see Section 3.4.5).

The lab setting in the traditional graphics state is used to control the number of tick marks on the axes. The setting is only used as a starting point for the algorithm R uses to determine sensible tick locations so the final number of tick marks that are drawn could easily differ from this specification. The setting takes two values: the first specifies the number of tick marks on the x-axis and the second specifies the number of tick marks on the y-axis.

The xaxp and yaxp settings also relate to the number and location of the tick marks on the axes of a plot. This setting is almost always calculated by R for each new plot so user settings are usually overridden (see Section 3.4.5 for an exception to this rule). In other words, it only makes sense to query this setting for its current value. The settings consist of three values: the first two specify the location of the left-most and right-most tick-marks (bottom and top tick-marks for the y-axis), and the third value specifies how many intervals there are between tick marks. When a log transformation is in effect for an axis, the three values have a different meaning altogether (see the on-line help page for par()).

The mgp setting controls the distance that the components of the axes are drawn away from the edge of the plot region. There are three values representing the positioning of the overall axis label, the tick mark labels, and the lines for the ticks. The values are in terms of lines of text away from the edges of the plot region. The default value is c(3, 1, 0). Figure 3.12 gives an example of different mgp settings.

The tck and tcl settings control the length of tick marks. The tcl setting

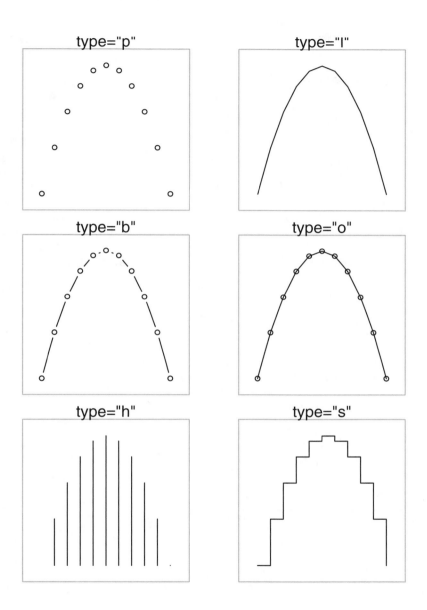

Figure 3.11
Basic plot types. Plotting the same data with different plot type settings. In each case, the output is produced by an expression of the form plot(x, y, type=*something*), where the relevant value of type is shown above each plot.

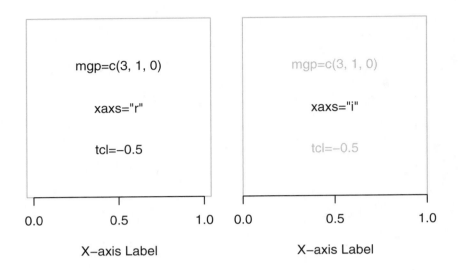

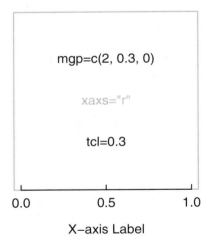

Figure 3.12
Different axis styles. The top-left plot demonstrates the default axis settings for
an x-axis. The top-right plot shows the effect of specifying an "internal" axis range
calculation and the bottom-left plot shows the effects of specifying different positions
for the axis labels and different lengths for the tick marks.

specifies the length of tick marks as a fraction of the height of a line of text. The sign dictates the direction of the tick marks — a negative value draws tick marks outside the plot region and a positive value draws tick marks inside the plot region. The `tck` setting specifies tick mark lengths as a fraction of the smaller of the physical width or height of the plotting region, but it is only used if its value is not `NA` (and it is `NA` by default). Figure 3.12 gives an example of different `tcl` settings.

The `xaxs` and `yaxs` settings control the "style" of the axes of a plot. By default, the setting is `"r"`, which means that R calculates the range of values on the axis to be wider than the range of the data being plotted (so that data symbols do not collide with the boundaries of the plot region). It is possible to make the range of values on the axis exactly match the range of values in the data, by specifying the value `"i"`. This can be useful if the range of values on the axes are being explicitly controlled via `xlim` or `ylim` arguments to a function. Figure 3.12 gives an example of different `xaxs` settings.

The `xaxt` and `yaxt` settings control the "type" of axes. The default value, `"s"`, means that the axis is drawn. Specifying a value of `"n"` means that the axis is not drawn.

The `xlog` and `ylog` settings control the transformation of values on the axes. The default value is `FALSE`, which means that the axes are linear and values are not transformed. If this value is `TRUE` then a logarithmic transformation is applied to any values on the relevant dimension in the plot region. This also affects the calculation of tick mark locations on the axes.

When data of a special nature are being plotted (e.g., time series data), some of these settings may not apply (and may not have any sensible interpretation).

The `bty` setting is not strictly to do with axes, but it controls the output of the `box()` function, which is most commonly used in conjunction with drawing axes. This function draws a bounding box around the edges of the plot region (by default). The `bty` setting controls the type of box that the `box()` function draws. The value can be `"n"`, which means that no box is drawn, or it can be one of `"o"`, `"l"`, `"7"`, `"c"`, `"u"`, or `"]"`, which means that the box drawn resembles the corresponding uppercase character. For example, `bty="c"` means that the bottom, left, and top borders will be drawn, but the right border will not be drawn.

In addition to these graphics state settings, many high-level plotting functions, e.g., `plot()`, provide arguments `xlim` and `ylim` to control the range of the scale on the axes. Section 2.2.2 has an example.

3.2.6   Plotting regions

As described in Section 3.1.1, the traditional graphics system defines several different regions on the graphics device. This section describes how to control the size and layout of these regions using graphics state settings. Figure 3.13 shows a diagram of some of the settings that affect the widths and horizontal placement of the regions.

Outer margins

By default, there are no outer margins on a page. Outer margins can be specified using the `oma` graphics state setting. This consists of four values for the four margins in the order (`bottom, left, top, right`) and values are interpreted as lines of text (a value of 1 provides space for one line of text in the margin). The margins can also be specified in inches using `omi` or in normalized device coordinates (i.e., as a proportion of the device region) using `omd`. In the latter case, the margins are specified in the order (`left, right, bottom, top`).

Figure regions

By default, the figure region is calculated from the settings for the outer margins, and the number of figures on the page. The figure region can be specified explicitly using either the `fig` setting or the `fin` state setting. The `fig` setting specifies the location, (`left, right, bottom, top`), of the figure region where each value is a proportion of the "inner" region (the page less the outer margins). The `fin` setting specifies the size, (`width, height`), of the figure region in inches and the resulting figure region is centered within the inner region.

Figure margins

The figure margins can be controlled using the `mar` state setting. This consists of four values for the four margins in the order (`bottom, left, top, right`) where each value represents a number of lines of text. The default is `c(5, 4, 4, 2) + 0.1`. The margins may also be specified in terms of inches using `mai`.

The `mex` setting controls the size of a "line" in the margins. This does not affect the size of text drawn in the margins, but is used to multiply the size of text to determine the height of one line of text in the margins.

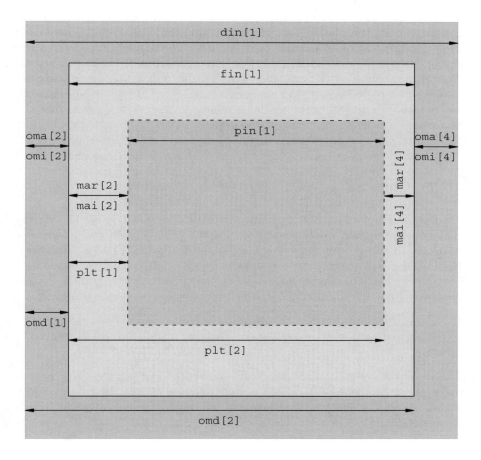

Figure 3.13
Graphics state settings controlling plot regions. These are some of the settings that
control the widths and horizontal locations of the plot regions. For ease of comparison, this diagram has the same layout as Figure 3.1: the central grey rectangle
represents the plot region, the lighter grey rectangle around that is the figure region,
and the darker grey rectangle around that is the outer margins. A similar diagram
could be produced for settings controlling heights and vertical locations.

Plot regions

By default, the plot region is calculated from the figure region less the figure margins. The location and size of the plot region may be controlled explicitly using the plt, pin, or pty settings. The plt setting allows the user to specify the location of the plot region, (left, right, bottom, top), where each value is a proportion of current figure region. The pin setting specifies the size of the plot region, (width, height), in terms of inches. The pty setting controls how much of the available space (figure region less figure margins) that the plot region occupies. The default value is "m", which means that the plot region occupies all of the available space. A value of "s" means that the plot region will take up as much of the available space as possible, but it must be "square" (i.e., its physical width will be the same as its physical height).

3.2.7 Clipping

Traditional graphics output is usually clipped to the plot region. This means that any output that would appear outside the plot region is not drawn. For example, in the default behavior, data symbols for (x, y) locations which lie outside the plot region are not drawn. Traditional graphics functions that draw in the margins clip output to the current figure region or to the device. Section 3.4 has information about which functions draw in which regions.

It can be useful to override the default clipping region. For example, this is necessary to draw a legend outside the plot region using the legend() function.

The traditional clipping region is controlled via the xpd setting. Clipping can occur either to the whole device (an xpd value of NA), to the current figure region (a value of TRUE), or to the current plot region (a value of FALSE, which is the default).

3.2.8 Moving to a new plot

As described in Section 2.1, high-level graphics functions usually start a new plot. There are traditional graphics state settings that control exactly when and how this happens.

The ask setting controls whether the user is prompted before the graphics system starts a new page of output. It is useful for viewing multiple pages of output (e.g., the output from example(boxplot)) that otherwise flick by too fast to view properly. If the ask setting is TRUE then the user is prompted before a new page of output is begun.

The **new** setting controls whether a function that starts a new plot will move on to the next figure region (possibly a new page). Every plot sets the value to **FALSE** so that the next plot will move on by default, but if this setting has the value **TRUE** then a new plot does not move on to the next figure region. This can be used to overlay several plots on the same figure (Section 3.4.7 has an example).

3.3 Arranging multiple plots

There are a number of ways to produce multiple plots on a single page.

The number of plots on a page, and their placement on the page, can be controlled directly by specifying traditional graphics state settings using the **par()** function, or through a higher-level interface provided by the **layout()** function. The **split.screen()** function (and associated functions) provide yet another approach where a figure region can itself be treated as a complete page to split into further figure and plot regions.

These three approaches are mutually incompatible. For example, a call to the **layout()** function will override any previous **mfrow** and **mfcol** settings. Also, some high-level functions (e.g., **coplot()**) call **layout()** or **par()** themselves to create a plot arrangement, which means that the output from such functions cannot be arranged with other plots on a page.

3.3.1 Using the traditional graphics state

The number of figure regions on a page can be controlled via the **mfrow** and **mfcol** graphics state settings. Both of these consist of two values indicating a number of rows, nr, and a number of columns, nc; these settings result in $nr \times nc$ figure regions of equal size.

The top-left figure region is used first. If the setting is made via **mfrow** then the figure regions along the top row are used next from left to right, until that row is full. After that, figure regions are used in the next row down, from left to right, and so on. When all rows are full, a new page is started. For example, the following code creates six figure regions on the page, arranged in three rows and two columns and the regions are used in the order shown in Figure 3.14a.

```
> par(mfrow=c(3, 2))
```

If the setting is made via `mfcol`, figure regions are used by column instead of by row.

The order in which figure regions are used can be controlled by using the `mfg` setting to specify the next figure region. This setting consists of two values that indicate the row and column of the next figure to use.

3.3.2 Layouts

The `layout()` function provides an alternative to the `mfrow` and `mfcol` settings. The primary difference is that the `layout()` function allows the creation of multiple figure regions of *unequal* sizes.

The simple idea underlying the `layout()` function is that it divides the inner region of the page into a number of rows and columns, but the heights of rows and the widths of columns can be independently controlled, *and* a figure can occupy more than one row or more than one column.*

The first argument (and the only required argument) to the `layout()` function is a matrix. The number of rows and columns in the matrix determines the number of rows and columns in the layout.

The contents of the matrix are integer values that determine which rows and columns each figure will occupy. The following layout specification is identical to `par(mfrow=c(3, 2))`.

```
> layout(matrix(c(1, 2, 3, 4, 5, 6), byrow=TRUE, ncol=2))
```

It may be easier to imagine the arrangement of figure regions if the matrix is specified using `cbind()` or `rbind()`. The code below repeats the previous example, but uses `rbind()` to specify the layout matrix.

```
> layout(rbind(c(1, 2),
               c(3, 4),
               c(5, 6)))
```

The function `layout.show()` may be helpful for visualizing the figure regions that are created. The following code creates a figure visualizing the layout created in the previous example (see Figure 3.14a).

```
> layout.show(6)
```

---

*The underlying concept of a "layout"[43] is also implemented, in a slightly different and more general way, in the grid graphics system (see Section 5.5.6)

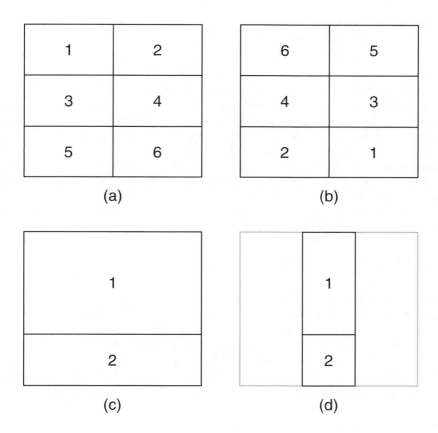

Figure 3.14

Some basic layouts. (a) A layout that is identical to `par(mfrow=c(3, 2))`. (b) Same as (a) except the figures are used in the reverse order. (c) A layout with unequal row heights. (d) same as (c) except the layout widths and heights "respect" each other.

The contents of the layout matrix determine the order in which the resulting figure regions will be used. The following code creates a layout with exactly the same rows and columns as the previous one, but the figure regions will be used in the reverse order (see Figure 3.14b).

```
> layout(rbind(c(6, 5),
               c(4, 3),
               c(2, 1)))
```

By default, all row heights are the same and all column widths are the same size and the available inner region is divided up equally. The `heights` arguments can be used to specify that certain rows are given a greater portion of the available height (for all of what follows, the `widths` argument works analogously for column widths). When the available height is divided up, the proportion of the available height given to each row is determined by dividing the row heights by the sum of the row heights. For example, in the following layout there are two rows and one column. The top row is given two-thirds of the available height $(2/(2 + 1))$ and the bottom row is given one third $(1/(2 + 1))$. Figure 3.14c shows the resulting layout.

```
> layout(matrix(c(1, 2)), heights=c(2, 1))
```

In the examples so far, the division of row heights has been completely independent of the division of column widths. The widths and heights can be forced to correspond as well so that, for example, a height of 1 corresponds to the same physical distance as a width of 1. This allows control over the aspect ratio of the resulting figure. The `respect` argument is used to force this correspondence. The following code is the same as the previous example except that the `respect` argument is set to `TRUE` (see Figure 3.14d).

```
> layout(matrix(c(1, 2)), heights=c(2, 1),
         respect=TRUE)
```

It is also possible to specify heights of rows and widths of columns in absolute terms. The `lcm()` function can be used to specify heights and widths for a layout in terms of centimeters. The following code is the same as the previous example, except that a third, empty, region is created to provide a vertical gap of 0.5cm between the two figures (see Figure 3.15a). The 0 in the first matrix argument means that no figure will ever occupy that region.

```
> layout(matrix(c(1, 0, 2)),
         heights=c(2, lcm(0.5), 1),
         respect=TRUE)
```

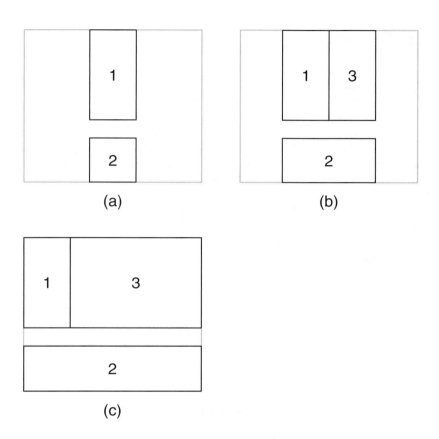

Figure 3.15
Some more complex layouts. (a) A layout with a row height specified in centimeters. (b) A layout with a figure occupying more than one column. (c) Same as (b), but with only column 1 and row 3 respected.

This next piece of code demonstrates that a figure may occupy more than one row or column in the layout. This extends the previous example by adding a second column and creating a figure region that occupies both columns of the bottom row. In the matrix argument, the value 2 appears in both columns of row 3 (see Figure 3.15b).

```
> layout(rbind(c(1, 3),
               c(0, 0),
               c(2, 2)),
         heights=c(2, lcm(0.5), 1),
         respect=TRUE)
```

Finally, it is possible to specify that only certain rows and columns should respect each other's heights/widths. This is done by specifying a matrix for the respect argument. In the following code, the previous example is modified by specifying that only the first column and the last row should respect each other's widths/heights. In this case, the effect is to ensure that the width of figure region 1 is the same as the height of figure region 2, but the width of figure region 3 is free to expand to the available width (see Figure 3.15c).

```
> layout(rbind(c(1, 3),
               c(0, 0),
               c(2, 2)),
         heights=c(2, lcm(0.5), 1),
         respect=rbind(c(0, 0),
                       c(0, 0),
                       c(1, 0)))
```

3.3.3   The split-screen approach

The split.screen() function provides yet another way to divide the page into a number of figure regions. The first argument, figs, is either two values specifying a number of rows and columns of figures (i.e., like the mfrow setting), or a matrix containing a figure region location, (left, right, bottom, top), on each row (i.e., like a fig setting on each row).

Having established figure regions in this manner, a figure region is used by calling the screen() function to select a region. This means that the order in which figures are used is completely under the user's control, and it is possible to reuse a figure region, though there are dangers in doing so (the on-line help for split.screen() discusses this some more). The function erase.screen() can be used to clear a defined screen and close.screen() can be used to remove one or more screen definitions.

An even more useful feature of this approach is that each figure region can itself be divided up by a further call to `split.screen()`. This allows complex arrangements of plots to be created.

The downside to this approach is that it does not fit very nicely with the underlying traditional graphics system model (see Section 3.1). The recommended way to achieve complex arrangements of plots is via the `layout()` function (see Section 3.3.2) or by using the grid graphics system (see Part II), possibly in combination with traditional high-level functions (see Appendix B).

3.4 Annotating plots

Sometimes it is not enough to be able to modify the default output from high-level functions and in many situations, further graphical output must be added to achieve the desired result (see, for example, Figure 1.3). R graphics in general is fundamentally oriented to supporting the annotation of plots — the ability to add graphical output to an existing plot. In particular, the regions and coordinate systems used in the construction of a plot are also available for adding further output to the plot. For example, it is possible to position a text label relative to the scales on the axes of a plot.

3.4.1 Annotating the plot region

Most graphics functions that add output to an existing plot, add the output to the plot region, relative to the user coordinate system.

Graphical primitives

This section describes the graphics functions that provide the most basic graphics output (lines, rectangles, text, etc).

The most common use of this facility is to plot additional sets of data within a plot. The `lines()` function draws lines between (`x, y`) locations, and the `points()` function draws data symbols at (`x, y`) locations. The following code demonstrates a common situation where three different sets of y-values, recorded at the same set of x-values, are plotted together on the same plot (see the top-left plot in Figure 3.16).

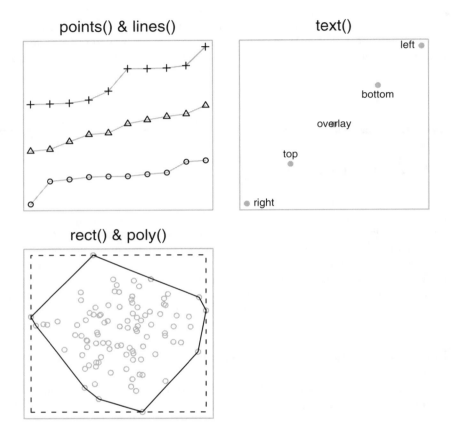

Figure 3.16
Annotating the plot region of a traditional graphics plot. The top-left plot shows points and extra lines being added to an initial line plot. The top-right plot shows text being added to an initial scatterplot. The bottom-left plot shows a dashed rectangle and a polygon being added to an initial scatterplot. Axes and labels have been omitted from the plots in order to avoid clutter.

First some data are generated, consisting of one set of x values and three sets of y values, and the first set of y values are plotted as a grey line (`type="1"` and `col="grey"`).

```
> x <- 1:10
> y <- matrix(sort(rnorm(30)), ncol=3)
> plot(x, y[,1], ylim=range(y), ann=FALSE, axes=FALSE,
        type="1", col="grey")
> box(col="grey")
```

Now a set of points are added for the first set of y values, then lines and points are added for the other two sets of y values.

```
> points(x, y[,1])
> lines(x, y[,2], col="grey")
> points(x, y[,2], pch=2)
> lines(x, y[,3], col="grey")
> points(x, y[,3], pch=3)
```

The `lines()` function typically draws a single line through many points (though `NA` values in the (x, y) locations will create breaks in the line). An alternative is provided by the `segments()` function, which will draw several different straight lines between pairs of end points.

It is also possible to draw text at (x, y) locations. This is useful for labeling data locations, particularly using the `pos` argument to offset the text so that it does not overlay any data symbols. The following code creates a diagram demonstrating the use of `text()` (see the top-right plot in Figure 3.16). Again, some data are created and (grey) data symbols are plotted at the (x, y) locations.

```
> x <- c(4, 5, 2, 1)
> y <- x
> plot(x, y, ann=FALSE, axes=FALSE, col="grey", pch=16)
> points(3, 3, col="grey", pch=16)
> box(col="grey")
```

Now some text labels are added, with each one offset in a different way from the (x, y) location. Notice that the arguments to `text()` may be vectors so that several pieces of text are drawn by the one function call.

```
> text(x, y, c("bottom", "left", "top", "right"), pos=1:4)
> text(3, 3, "overlay")
```

There are also the functions `rect()` and `polygon()` for drawing rectangles and polygons. The arguments to `rect()` may be vectors, in which case multiple rectangles are drawn. Multiple polygons may be drawn using `polygon()` by inserting an `NA` value between each set of polygon vertexes. R will draw self-intersecting polygons, but does not handle polygons with holes. For both `rect()` and `polygon()`, the `col` argument specifies the color to *fill* the interior of the shape and the argument `border` controls the color of the line around the boundary of the shape. The following code demonstrates the use of these functions. First, data are generated and plotted (as grey circles).

```
> x <- rnorm(100)
> y <- rnorm(100)
> plot(x, y, ann=FALSE, axes=FALSE, col="grey")
> box(col="grey")
```

Now we draw a dashed bounding box for the data using `rect()` and a solid convex hull using `polygon()` (and `chull()` to calculate the hull; see the bottom-left plot of Figure 3.16).

```
> rect(min(x), min(y), max(x), max(y), lty="dashed")
> hull <- chull(x, y)
> polygon(x[hull], y[hull])
```

Like the `plot()` function, the `text()`, `lines()`, and `points()` functions are generic. This means that they have flexible interfaces for specifying (x, y) locations, or they produce different output when given objects of a particular class in the `x` argument. For example, both `lines()`, and `points()` will accept formulae for specifying the (x, y) locations and the `lines()` function will behave sensibly when given a `ts` (time series) object to draw.

As a parallel to the `matplot()` function (see page 29), there are functions `matpoints()` and `matlines()` specifically for adding lines and data symbols to a plot given x or y as matrices.

Graphical utilities

In addition to the low-level graphical primitives of the previous section, there are a number of utility functions that provide a set of slightly more complex shapes.

The `grid()` function adds a series of grid lines to a plot. This is simply a series of line segments, but the default appearance (light grey and dotted) is suited to the purpose of providing visual cues to the viewer without interfering with the primary data symbols.

abline() & arrows()

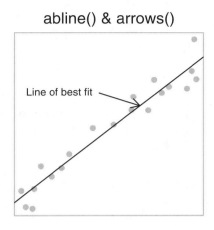

rug()

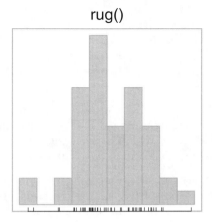

Figure 3.17
More examples of annotating the plot region of a traditional graphics plot. The left-hand plot shows a line of best fit (plus a text label and arrow) being added to an initial scatterplot. The right-hand plot shows a series of ticks being added as a rug plot on an initial histogram.

The `abline()` function provides a number of convenient ways to add a line (or lines) to a plot. The line(s) can be specified either by a slope and y-axis intercept, or as a series of x-locations for vertical lines or y-locations for horizontal lines. The function will also accept the coefficients from a linear regression analysis (even as an `lm` object), thereby providing a simple way to add a line of best fit to a scatterplot.

The `arrows()` function draws line segments and augments them with simple arrowheads at either end. The following code annotates a basic scatterplot with a line and arrows (see the left plot of Figure 3.17).

First, some data are generated and plotted.

```
> x <- runif(20, 1, 10)
> y <- x + rnorm(20)
> plot(x, y, ann=FALSE, axes=FALSE, col="grey", pch=16)
> box(col="grey")
```

Now a line of best fit is drawn through the data using `abline()` and a text label and arrow are added using `text()` and `arrows()`.

```
> lmfit <- lm(y ~ x)
> abline(lmfit)
> arrows(5, 8, 7, predict(lmfit, data.frame(x=7)),
         length=0.1)
> text(5, 8, "Line of best fit", pos=2)
```

The box() function draws a rectangle around the boundary of the plot region. The which argument makes it possible to draw the rectangle around the current figure region, inner region, or outer region instead. The box() function has been used in each of the examples in this section.

The rug() function produces a "rug" plot along one of the axes, which consists of a series of tick marks representing data locations. This can be useful to represent an additional one-dimensional plot of data (e.g., in combination with a density curve). The following code uses this function to annotate a histogram (see the right plot of Figure 3.17).

```
> y <- rnorm(50)
> hist(y, main="", xlab="", ylab="", axes=FALSE,
       border="grey", col="light grey")
> box(col="grey")
> rug(y, ticksize=0.02)
```

3.4.2 Missing values and non-finite values

R has special values representing missing observations (NA) and non-finite values (NaN and Inf). Most traditional graphics functions allow such values within (x, y) locations and handle them by not drawing the relevant location. For drawing data symbols or text, this means the relevant data symbol or piece of text will not be drawn. For drawing lines, this means that lines to or from the relevant location are not drawn; a gap is created in the line. For drawing rectangles, an entire rectangle will not be drawn if any of the four boundary locations are missing or non-finite.

Polygons are a slightly more complex case. For drawing polygons, a missing or non-finite value in x or y is interpreted as the end of one polygon and the start of another. Figure 3.18 shows an example. On the left, a polygon is drawn through 12 locations evenly spaced around a circle. On the right, the first, fifth, and ninth locations have been set to NA so the output is split into three separate polygons.

Missing or non-finite values can also be specified for some traditional graphics state settings. For example, if a color setting is missing or non-finite then

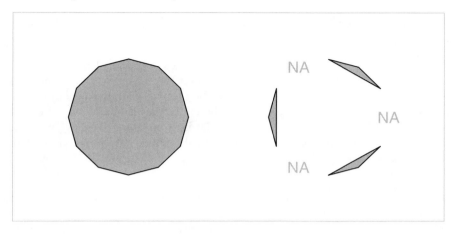

Figure 3.18
Drawing polygons using the `polygon()` function. On the left, a single polygon
(dodecagon) is produced from multiple (`x`, `y`) locations. On the right, the first,
fifth, and ninth values have been set to `NA`, which splits the output into three separate
polygons.

nothing is drawn (this is a brute-force way to specify a completely transparent
color). Similarly, specifying a missing value or non-finite value for `cex` means
that the relevant data symbol or piece of text is not drawn.

3.4.3   Annotating the margins

There are only two functions that produce output in the figure or outer margins, relative to the margin coordinate systems (Section 3.1.1).

The `mtext()` function draws text at any location in any of the margins. The
`outer` argument controls whether output goes in the figure or outer margins.
The `side` argument determines which margin to draw in: 1 means the bottom
margin, 2 means the left margin, 3 means the top margin, and 4 means the
right margin.

Text is drawn a number of lines of text away from the edges of the plot region
for figure margins, or a number of lines away from the edges of the inner region
for outer margins. In the figure margins, the location of the text along the
margin can be specified relative to the user coordinates on the relevant axis
using the `at` argument. In some cases it is possible to specify the location
as a proportion of the length of the margin using the `adj` argument, but this
is dependent on the value of the `las` state setting. For certain `las` settings,
the `adj` argument instead controls the justification of the text relative to a

position chosen by the `las` argument. Often, a trial-and-error approach is required to achieve the desired result.

The `title()` function is essentially a specialized version of `mtext()`. It is more convenient for producing a few specific types of output, but much less flexible than `mtext()`. This function can be used to produce a main title for a plot (in the top figure margin), axis labels (in the left and bottom figure margins), and a sub-title for a plot (in the bottom margin below the x-axis label). The output from this function is heavily influenced by various graphics state settings, such as `cex.main` and `col.main` (for the size and color of the title).

With a little extra effort, it is also possible to produce graphical output in the figure or outer margins using the functions that normally draw in the plot region (e.g., `points()` and `lines()`). In order to do this, the clipping region of the plot must first be set using the `xpd` state setting (see Section 3.2.7). This approach is not very convenient because the functions are drawing relative to user coordinates rather than locations relative to the margin coordinate systems. Nevertheless, it can sometimes be useful.

The following code demonstrates the use of `mtext()` and a simple application of using `lines()` outside the plot region for drawing what appears to be a rectangle extending across two plots (see Figure 3.19).*

First of all, the `mfrow` setting is used to set up an arrangement of two figure regions, one above the other. The clipping region is set to the entire device using `xpd=NA`.

```
> y1 <- rnorm(100)
> y2 <- rnorm(100)
```

```
> par(mfrow=c(2, 1), xpd=NA)
```

The first data set is plotted as a time series on the top plot and a label is added at the left end of figure margin 3. In addition, thick grey lines are drawn to represent the top of the rectangle that deliberately extend well below the bottom of the plot.

---

*This example was motivated by a question to R-help on December 14 2004 with subject: "drawing a rectangle through multiple plots".

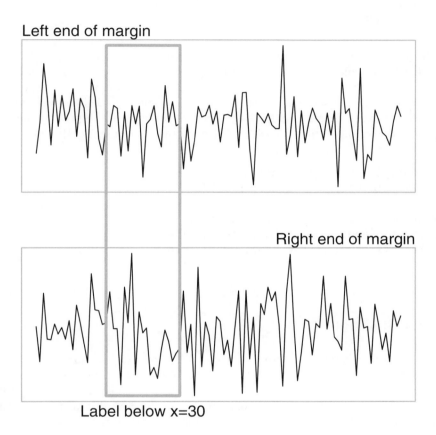

Figure 3.19
Annotating the margins of a traditional graphics plot. Text has been added in
margin 3 of the top plot and in margins 1 and 3 in the bottom plot. Thick grey
lines have been added to both plots (and overlapped so that it appears to be a single
rectangle across the plots).

```
> plot(y1, type="l", axes=FALSE,
       xlab="", ylab="", main="")
> box(col="grey")
> mtext("Left end of margin", adj=0, side=3)
> lines(x=c(20, 20, 40, 40), y=c(-7, max(y1), max(y1), -7),
        lwd=3, col="grey")
```

The second data set is plotted as a time series in the bottom plot, a label is
added to this plot at the right end of figure margin 3, and another label is
drawn beneath the x-location 30 in figure margin 1. Finally, thick grey lines
are drawn to represent the bottom of the rectangle that deliberately extend
above the plot. These lines overlap the lines drawn with respect to the top
plot to create the impression of a single rectangle traversing both plots.

```
> plot(y2, type="l", axes=FALSE,
       xlab="", ylab="", main="")
> box(col="grey")
> mtext("Right end of margin", adj=1, side=3)
> mtext("Label below x=30", at=30, side=1)
> lines(x=c(20, 20, 40, 40), y=c(7, min(y2), min(y2), 7),
        lwd=3, col="grey")
```

3.4.4   Legends

The traditional graphics system provides the `legend()` function for adding a
legend or key to a plot. The legend is usually drawn within the plot region,
and is located relative to user coordinates. The function has many arguments,
which allow for a great deal of flexibility in the specification of the contents
and layout of the legend. The following code demonstrates a couple of typical
uses.

The first example shows a scatterplot with a legend to relate group names to
different symbols (see the top plot in Figure 3.20).

```
> with(iris,
       plot(Sepal.Length, Sepal.Width,
            pch=as.numeric(Species), cex=1.2))
> legend(6.1, 4.4, c("setosa", "versicolor", "virginica"),
         cex=1.5, pch=1:3)
```

The next example shows a barplot with a legend to relate group names to

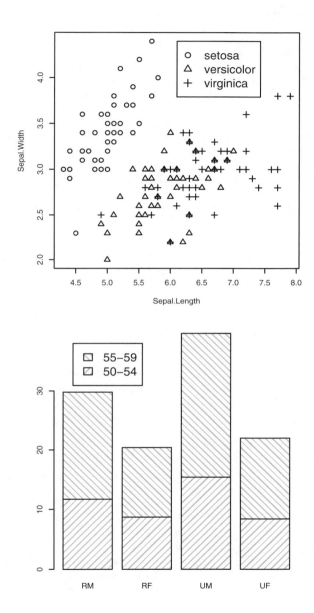

Figure 3.20
Some simple legends. Legends can be added to any kind of plot and can relate text labels to different symbols or different fill colors or patterns.

different fill patterns (see the bottom plot in Figure 3.20).*

```
> barplot(VADeaths[1:2,], angle=c(45, 135), density=20,
         col="grey", names=c("RM", "RF", "UM", "UF"))
> legend(0.4, 38, c("55-59", "50-54"), cex=1.5,
         angle = c(135, 45), density = 20, fill = "grey")
```

It should be noted that it is entirely the responsibility of the user to ensure that the legend corresponds to the plot. There is no automatic checking that data symbols in the legend match those in the plot, or that the labels in the legend have any correspondence with the data.

Some high-level functions draw their own legend specific to their purpose (e.g., `filled.contour()`).

3.4.5   Axes

In most cases, the axes that are automatically generated by the traditional graphics system will be sufficient for a plot. This is true even when the data being plotted on an axis are non-numeric. For example, the axes of a boxplot or barplot are labeled appropriately using group names.

Section 3.2.5 describes ways in which the default appearance of automatically-generated axes can be modified, but it is more often the case that the user needs to inhibit the production of the automatic axis and draw a customized axis using the `axis()` function.

The first step is to inhibit the default axes. Most high-level functions should provide an `axes` argument which, when set to `FALSE`, indicates that the high-level function should not draw axes. Specifying the traditional graphics setting `xaxt="n"` (or `yaxt="n"`) may also do the trick.

The `axis()` function can draw axes on any side of a plot (chosen by the `side` argument), and the user can specify the location along the axis of tick marks and the text to use for tick labels (using the `at` and `labels` arguments respectively). The following code demonstrates a simple example of a plot where the automatic axes are inhibited and custom axes are drawn, including a "secondary" y-axis on the right side of the plot (see Figure 3.21).

First of all, some temperature data are generated and an empty plot is created with no data symbols and no axes.

---

*The data for the scatterplot are from the `iris` data set (see page 29) and the data for the histogram are from the `VADeaths` data set (see page 3).

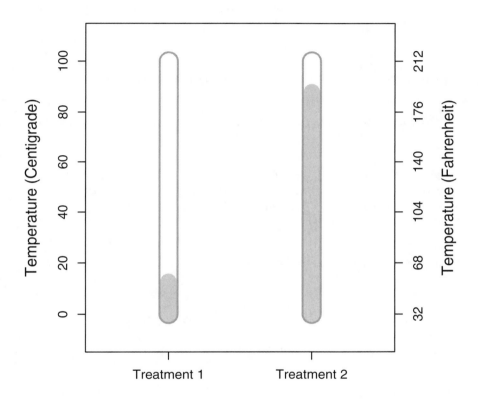

Figure 3.21
Customizing axes. An initial plot is drawn with a y-scale in degrees centigrade, then
a secondary y-axis is drawn with a scale in degrees Fahrenheit. The x-axis is drawn
using special text labels, rather than the default numeric locations of the tick marks.

```
> x <- 1:2
> y <- runif(2, 0, 100)
> par(mar=c(4, 4, 2, 4))
> plot(x, y, type="n", xlim=c(0.5, 2.5), ylim=c(-10, 110),
        axes=FALSE, ann=FALSE)
```

Next, the main y-axis is drawn with specific tick locations to represent the Centigrade scale.

```
> axis(2, at=seq(0, 100, 20))
> mtext("Temperature (Centigrade)", side=2, line=3)
```

Now the bottom axis is drawn with special labels and a secondary y-axis is drawn to represent the Fahrenheit scale.

```
> axis(1, at=1:2, labels=c("Treatment 1", "Treatment 2"))
> axis(4, at=seq(0, 100, 20), labels=seq(0, 100, 20)*9/5 + 32)
> mtext("Temperature (Fahrenheit)", side=4, line=3)
> box()
```

Finally, some thermometer-like symbols are drawn to represent the actual temperatures.

```
> segments(x, 0, x, 100, lwd=20, col="dark grey")
> segments(x, 0, x, 100, lwd=16, col="white")
> segments(x, 0, x, y, lwd=16, col="light grey")
```

The axis() function is not generic, but there are special alternative functions for plotting time related data. The functions axis.Date() and axis.POSIXct() take an object containing dates and produce an axis with appropriate labels representing times, days, months, and years (e.g., 10:15, Jan 12 or 1995).

In some cases, it may be useful to draw tick marks at the locations that the default axis would use, but with different labels. The axTicks() function can be used to calculate these default locations. This function is also useful for enforcing an xaxp (or yaxp) graphics state setting. If these settings are specified via par(), they usually have no effect because the traditional graphics system almost always calculates the settings itself. The user can choose these settings by passing them as arguments to axTicks(), then passing the resulting locations via the at argument to axis().

3.4.6 Mathematical formulae

Any R graphics function that draws text should accept both a normal string, e.g., `"some text"`, and an R expression, which is typically the result of a call to the `expression()` function. If an expression is specified as the text to draw, then it is interpreted as a mathematical formula and is formatted appropriately. This section provides some simple examples of what can be achieved. For a complete description of the available features, type `help(plotmath)` or `demo(plotmath)` in an R session.[*]

When an R expression is provided as text to draw in graphical output, the expression is evaluated to produce a mathematical formula. This evaluation is very different from the normal evaluation of R expressions: certain names are interpreted as special mathematical symbols, e.g., `alpha` is interpreted as the Greek symbol α; certain mathematical operators are interpreted as literal symbols, e.g., a `+` is interpreted as a plus sign symbol; and certain functions are interpreted as mathematical operators, e.g., `sum(x, i==1, n)` is interpreted as $\sum_{i=1}^{n} x$. Figure 3.22 shows some examples of expressions and the output that they create.

In some situations, for example, when calling graphics functions from within a loop, or when calling graphics functions from within another function, the expression representing the mathematical formula must be constructed using values within variables as well as literal symbols and constants. A variable name within an expression will be treated as a literal symbol (i.e., the variable name will be drawn, not the value within the variable). The solution in such cases is to use the `substitute()` function to produce an expression. The following code shows the use of `substitute()` to produce a label where the year is stored in a variable.

```
> myfunction <- function(year) {
    text(0.5, 0.5, substitute(paste("Temperature (",
                              degree, "C) in ", year),
        list(year=year)))
  }
```

The mathematical annotation feature makes use of information about the dimensions of individual characters to perform the formatting of the formula. For some output formats, such information is not available, so mathematical formulae cannot be produced. However, mathematical formulae are supported on the major screen devices (X11, Windows, and Quartz) and information

---

[*]Further information can also be obtained from an article in the Journal of Computational and Graphical Statistics[45].

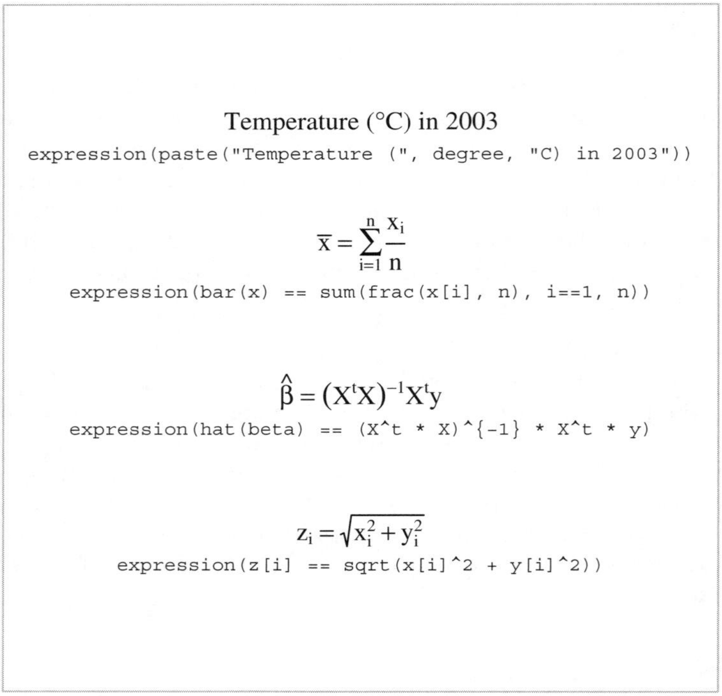

Temperature (°C) in 2003

```
expression(paste("Temperature (", degree, "C) in 2003"))
```

$$\overline{x} = \sum_{i=1}^{n} \frac{x_i}{n}$$

```
expression(bar(x) == sum(frac(x[i], n), i==1, n))
```

$$\hat{\beta} = (X^t X)^{-1} X^t y$$

```
expression(hat(beta) == (X^t * X)^{-1} * X^t * y)
```

$$z_i = \sqrt{x_i^2 + y_i^2}$$

```
expression(z[i] == sqrt(x[i]^2 + y[i]^2))
```

Figure 3.22
Mathematical formulae in plots. For each example, the output is shown in a serif font, and below that, in a typewriter font, is the R expression required to produce the output.

for the standard Adobe Type 1 fonts is distributed with R so mathematical formulae should always be available for PostScript and PDF output.

3.4.7 Coordinate systems

The traditional graphics system provides a number of coordinate systems for conveniently locating graphical output (see Section 3.1.1). Graphical output in the plot region is automatically positioned relative to the scales on the axes and text in the figure margins is placed in terms of a number of lines away from the edge of the plot (i.e., a scale that naturally corresponds to the size of the text).

It is also possible to locate output according to other coordinate systems that are not automatically supplied, but a little more work is required by the user. The basic principle is that the traditional graphics state can be queried to determine features of existing coordinate systems, then new coordinate systems can be calculated from this information.

The par function

As well as being used to enforce new graphics state settings, the function `par()` can also be used to query current graphics state settings. The most useful settings are: `din`, `fin`, and `pin`, which reflect the current size, (`width`, `height`), of the graphics device, figure region, and plot region, in inches; and `usr`, which reflects the current user coordinate system (i.e., the ranges on the axes). The values of `usr` are in the order (`xmin, xmax, ymin, ymax`). When a scale has a logarithmic transformation, the values are (`10^xmin, 10^xmax, 10^ymin, 10^ymax`).

There are also settings that reflect the size, (`width, height`), of a "standard" character. The setting `cin` gives the size in inches, `cra` in "rasters" or pixels, and `cxy` in "user coordinates." However, these values are not very useful because they only refer to a `cex` value of 1 (i.e., they ignore the current `cex` setting) *and* they only refer to the `ps` value when the current graphics device was first opened. Of more use are the `strheight()` function and the `strwidth()` function. These calculate the height and width of a given piece of text in inches, or in terms of user coordinates, or as a proportion of the current figure region (taking into account the current `cex` and `ps` settings).

The following code demonstrates a simple example of making use of customized coordinates where a ruler is drawn showing centimeter units (see Figure 3.23).

A blank plot region is set up first and calculations are performed to establish

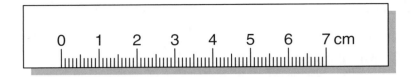

Figure 3.23
Custom coordinate systems. The lines and text are drawn relative to real physical centimeters (rather than the default coordinate system defined by the scales on plot axes).

the relationship between user coordinates in the plot and physical centimeters.*

```
> plot(0:1, 0:1, type="n", axes=FALSE, ann=FALSE)
> usr <- par("usr")
> pin <- par("pin")
> xcm <- diff(usr[1:2])/(pin[1]*2.54)
> ycm <- diff(usr[3:4])/(pin[2]*2.54)
```

Now drawing can occur with positions expressed in terms of centimeters. First of all a "drop shadow" is drawn to give a three-dimensional effect by drawing a grey rectangle offset by 2mm from the main ruler. The call to **par()** makes sure that the grey rectangle is not clipped to the plotting region (see Section 3.2.7).

```
> par(xpd=NA)
> rect(0 + 0.2*xcm, 0 - 0.2*ycm,
       1 + 0.2*xcm, 1 - 0.2*ycm,
       col="grey", border=NA)
```

The ruler itself is drawn with a call to **rect()** to draw the edges of the ruler, a call to **segments()** to draw the scale, and calls to **text()** to label the scale.

---

*R graphics relies on a graphics device providing accurate information on the physical size of the natural units on the device (e.g., the physical size of pixels on a computer screen). If a graphics device does not give accurate information, when R attempts to draw output with an physical size (e.g., a line 1 inch long), it may not appear with the exact physical size on the device. The physical size of output for PostScript and PDF files should always be correct, but small inaccuracies may occur when specifying output with an physical size (such as inches) on screen devices such as Windows and X11 windows.

```
> rect(0, 0, 1, 1, col="white")
> segments(seq(1, 8, 0.1)*xcm, 0,
           seq(1, 8, 0.1)*xcm,
           c(rep(c(0.5, rep(0.25, 4),
                   0.35, rep(0.25, 4)),
                 7), 0.5)*ycm)
> text(1:8*xcm, 0.6*ycm, 0:7, adj=c(0.5, 0))
> text(8.2*xcm, 0.6*ycm, "cm", adj=c(0, 0))
```

There are utility functions, xinch() and yinch(), for performing the inches-to-user coordinates transformation (plus xyinch() for converting a location in one step and cm() for converting inches to centimeters).

One problem with performing coordinate transformations like these is that the locations and sizes being drawn have no memory of how they were calculated. They are specified as locations and dimensions in user coordinates. This means that if the device is resized (so that the relationship between physical dimensions and user coordinates changes), the locations and sizes will no longer have their intended meaning. If, in the above example, the device is resized, the ruler will no longer accurately represent centimeter units. This problem will also occur if output is copied from one device to another device that has different physical dimensions. The legend() function performs calculations like these when arranging the components of a legend and its output is affected by device resizes and copying between devices.*

Overlaying output

It is sometimes useful to plot two data sets on the same plot where the data sets share a common x-variable, but have very different y-scales. This can be achieved in at least two ways. One approach is simply to use par(new=TRUE) to overlay two distinct plots on top of each other (care must be taken to avoid conflicting axes overwriting each other). Another approach is to explicitly reset the usr state setting before plotting a second set of data. The following code demonstrates both approaches to produce exactly the same result (see the top plot of Figure 3.24).

The data are yearly numbers of drunkenness-related arrests[†] and mean annual temperature in New Haven, Connecticut from 1912 to 1971. The temperature

---

*It is possible to work around these problems in R version 2.1.0 and above by using the recordGraphics() function, although this function should be used with extreme care.

[†]These data were obtained from "Crime Statistics and Department Demographics" on the New Haven Police Department Web Site:
http://www.cityofnewhaven.com/police/html/stats/crime/yearly/1863-1920.htm

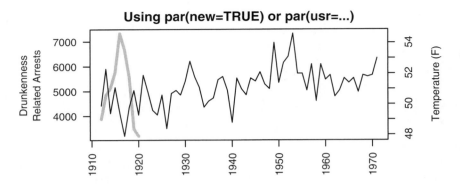

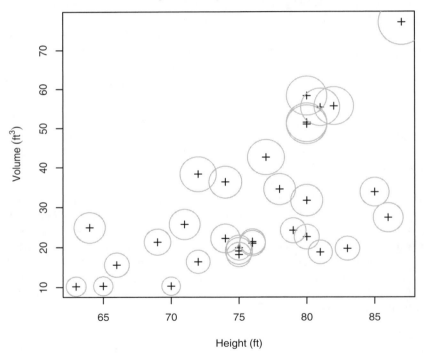

Figure 3.24

Overlaying plots. In the top plot, two line plots are drawn one on top of each other
to produce aligned plots of two data sets with very different scales. In the bottom
plot, the plotting function `symbols()` is used in "annotating mode" so that it adds
circles to an existing scatterplot rather than producing a complete plot itself.

data are available as the data set `nhtemp` in the `datasets` package. There are only arrests data for the first 9 years.

```
> drunkenness <- ts(c(3875, 4846, 5128, 5773, 7327,
                      6688, 5582, 3473, 3186,
                      rep(NA, 51)),
                    start=1912, end=1971)
```

The first approach is to draw a plot of the drunkenness data, call `par(new=TRUE)`, then draw a complete second plot of the temperature data on top of the first plot. The second plot does not draw default axes (`axes=FALSE`), but uses the `axis()` function to draw a secondary y-axis to represent the temperature scale.

```
> par(mar=c(5, 6, 2, 4))
> plot(drunkenness, lwd=3, col="grey", ann=FALSE, las=2)
> mtext("Drunkenness\nRelated Arrests", side=2, line=3.5)
> par(new=TRUE)
> plot(nhtemp, ann=FALSE, axes=FALSE)
> mtext("Temperature (F)", side=4, line=3)
> title("Using par(new=TRUE)")
> axis(4)
```

The second approach draws only one plot (for the drunkenness data). The user coordinate system is then redefined by specifying a new `usr` setting and the second "plot" is produced simply using `lines()`. Again, a secondary axis is drawn using the `axis()` function.

```
> par(mar=c(5, 6, 2, 4))
> plot(drunkenness, lwd=3, col="grey", ann=FALSE, las=2)
> mtext("Drunkenness\nRelated Arrests", side=2, line=3.5)
> usr <- par("usr")
> par(usr=c(usr[1:2], 47.6, 54.9))
> lines(nhtemp)
> mtext("Temperature (F)", side=4, line=3)
> title("Using par(usr=...)")
> axis(4)
```

Some high-level functions (e.g., `symbols()` and `contour()`) provide an argument called `add` which, if set to `TRUE`, will add the function output to the current plot, rather than starting a new plot. The following code shows the `symbols()` function being used to annotate a basic scatterplot (see the bottom plot of Figure 3.24). The data used in this example are from the `trees` data set (see page 35).

```
> with(trees,
      {
         plot(Height, Volume, pch=3,
               xlab="Height (ft)",
               ylab=expression(paste("Volume ", (ft^3))))
         symbols(Height, Volume, circles=Girth/12,
                  fg="grey", inches=FALSE, add=TRUE)
      })
```

Another function of this type is the `bxp()` function. This function is called by `boxplot()` to draw the individual boxplots and is specifically set up to add boxplots to an existing plot (although it can also produce a complete plot).

It is also worth remembering that R follows a painters model, with later output obscuring earlier output. The following example makes use of this feature to fill a complex region within a plot (see Figure 3.25).

The first step is to generate some data and calculate some important features of the data.

```
> xx <- c(1:50)
> yy <- rnorm(50)
> n <- 50
> hline <- 0
```

The first thing to draw is a plot with a filled polygon beneath the y-values (see the top-left plot of Figure 3.25).

```
> plot (yy ~ xx, type="n", axes=FALSE, ann=FALSE)
> polygon(c(xx[1], xx, xx[n]), c(min(yy), yy, min(yy)),
          col="grey", border=NA)
```

The next step is to draw a rectangle over the top of the polygon up to a fixed y-value. The expression `par("usr")` is used to obtain the current x-scale and y-scale ranges (see the top-right plot of Figure 3.25).

```
> usr <- par("usr")
> rect(usr[1], usr[3], usr[2], hline, col="white", border=NA)
```

Now a line through the y-values is drawn over the top of the rectangle (see the bottom-left plot of Figure 3.25).

```
> lines(xx, yy)
```

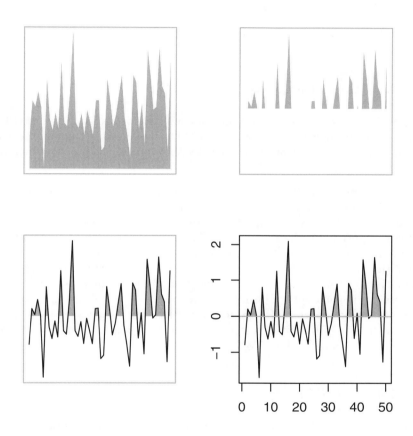

Figure 3.25
Overlaying output (making use of the painters model). The final complex plot, shown at bottom-right, is the result of overlaying several basic pieces of output: a grey polygon at top-left, with a white rectangle over the top (top-right), a black line on top of that (bottom-left), and a grey line on top of it all (plus axes and a bounding box).

Finally, a horizontal line is drawn to indicate the y-value cut-off, and axes are added to the plot (see the bottom-right plot of Figure 3.25).

```
> abline (h=hline,col="grey")
> box()
> axis(1)
> axis(2)
```

3.4.8 Bitmap images

The R graphics engine has no internal support for drawing bitmaps. Despite this, bitmap images can be represented by drawing a rectangle for each pixel in the image. A convenient interface for this approach is provided by functions in the `pixmap` package[8].

The plot in Figure 3.26 shows an example of what can be achieved using the functions in the `pixmap` package. This plot shows the relationship between the time of day that every second low tide occurred and the phase of the moon, for the port of Auckland, New Zealand in February 2005. The `addlogo()` function has been used to add a bitmap of the moon as a dramatic backdrop for the main plot (the code is not shown, but it is available on the web site for this book). This approach is most appropriate for producing images on screen or in some sort of bitmap format such as PNG. When used for creating vector formats such as PostScript and PDF, the file size grows very rapidly with the size of the bitmap (e.g., the PostScript file for the printed version of Figure 3.26 is more than 5MB!).

3.4.9 Special cases

Some high-level functions are a little more difficult to annotate than others because the plotting regions that they set up either are not immediately obvious, or are not available after the function has run. This section describes a number of high-level functions where additional knowledge is required to perform annotations.

Obscure scales on axes

It is not immediately obvious how to add extra annotation to a barplot or a boxplot in traditional R graphics because the scale on the categorical axis is not obvious.

The difficulty with the `barplot()` function is that because the scale on the

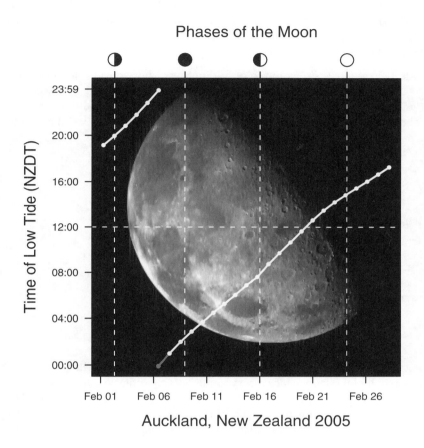

Figure 3.26

Adding a bitmap to a plot. A plot with a bitmap of the moon as a back-drop, added using the `pixmap` package. The bitmap is a view of the Moon's north pole assembled from images taken by the Galileo spacecraft, courtesy of NASA (`image #: PIA00130`). The data on low tides and phases of the moon for Auckland in February 2005 were obtained from Land Information New Zealand (`http://hydro.linz.govt.nz`).

x-axis is not labelled at all by default. the numeric scale is not obvious (and calling `par("usr")` is not much help because the scale that the function sets up is not intuitive either). In order to add annotations sensibly to a barplot it is necessary to capture the value returned by the function. This return value gives the x-locations of the mid-points of each bar that the function has drawn. These midpoints can then be used to locate annotations relative to the bars in the plot.

The code below shows an example of adding extra horizontal reference lines to the bars of a barplot. The mid-points of the bars are saved to a variable called `midpts`, then locations are calculated from those mid-points (and the original counts) to draw horizontal white line segments within each bar using the `segments()` function (see the left plot of Figure 3.27).

```
> y <- sample(1:10)
> midpts <- barplot(y, col=" light grey")
> width <- diff(midpts[1:2])/4
> left <- rep(midpts, y - 1) - width
> right <- rep(midpts, y - 1) + width
> heights <- unlist(apply(matrix(y, ncol=10),
                          2, seq))[-cumsum(y)]
> segments(left, heights, right, heights,
           col="white")
```

The `boxplot()` function is similar to the `barplot()` function in that the x-scale is typically labelled with category names so the numeric scale is not obvious from looking at the plot. Fortunately, the scale set up by the `boxplot()` function is much more intuitive. The individual boxplots are drawn at x-locations `1:n`, where `n` is the number of boxplots being drawn.

The following code provides a simple example of annotating boxplots to add a jittered dotplot of individual data points on top of the boxplots. This provides a detailed view of the data as well as showing the main features via the boxplot. It is also a useful way to show how interesting features of the data, such as small clusters of points, can be hidden by a boxplot. In this example, the jittered data are centered upon the x-locations `1:2` to correspond to the centers of the relevant boxplots (see the right plot of Figure 3.27).[*]

---

[*]The data used in this example are from the `ToothGrowth` data set (see page 3).

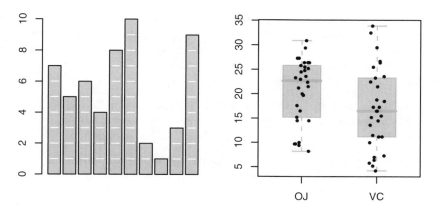

Figure 3.27
Special-case annotations. Some examples of functions where annotation requires special care. In the barplot at left, the value returned by the `barplot()` function is used to add horizontal white lines within the bars. Jittered points are added to the boxplot (right) using the knowledge that the *i*th box is located at position *i* on the x-axis.

```
> with(ToothGrowth,
      {
        boxplot(len ~ supp, border="grey",
                col="light grey", boxwex=0.5)
        points(jitter(rep(1:2, each=30), 0.5),
               unlist(split(len, supp)),
               cex=0.5, pch=16)
      })
```

Functions that draw several plots

The `pairs()` function is an example of a high-level function that draws more than one plot. This function draws a matrix of scatterplots. Such functions tend to save the traditional graphics state before drawing, call `par(mfrow)` or `layout()` to arrange the individual plots, and restore the traditional graphics state once all of the individual plots have been drawn. This means that it is not possible to annotate any of the plots drawn by the `pairs()` function once the function has completed drawing. The regions and coordinate systems that the function set up to draw the individual plots have been thrown away. The only way to annotate the output from such functions is by way of "panel" functions.

The `pairs()` function has a number of arguments that allow the user to specify a function: `panel`, `diag.panel`, `upper.panel`, `lower.panel`, and `text.panel`. The functions specified via these arguments are run as each individual plot is drawn. In this way, the panel function has access to the plot regions that are set up for each individual plot.

The `filled.contour()` function and the `coplot()` function have the same problem as `pairs()`, as the legends that they draw are actually separate plots. Again, they allow annotation via panel function arguments.

The following code demonstrates a simple use of a panel function with the `coplot()` function. The main conditioning plot shows the locations of earthquakes in the Pacific Ocean near Fiji since 1964,* available as the `quakes` data set in the `datasets` package. There are multiple panels, each of which shows the earthquakes that occurred at a particular range of depths. A panel function is specified via the `panel` argument to add maps of Fiji and New Zealand to each panel of `coplot()` output (see Figure 3.28).

The panel function first calls the `rect()` function to overlay a white background and hide the default grid lines. Next, the panel function calls the `points()` function to draw the points that would normally be drawn, but uses a custom plotting symbol (a very small dot). The `map()` function is called to draw the maps of Fiji and the top of the North Island of New Zealand, and the `text()` function is used to add country names. The map is drawn using the `map()` function from the `maps` package.

```
> library(maps)
> coplot(lat ~ long | depth, data = quakes, number=4,
        panel=function(x, y, ...) {
            usr <- par("usr")
            rect(usr[1], usr[3], usr[2], usr[4], col="white")
            map("world2", regions=c("New Zealand", "Fiji"),
                add=TRUE, lwd=0.1, fill=TRUE, col="grey")
            text(180, -13, "Fiji", adj=1, cex=0.7)
            text(170, -35, "NZ", cex=0.7)
            points(x, y, pch=".")
        })
```

There is a predefined panel function called `panel.smooth()`, which draws points and then adds a smoothed line through the points.

---

*The data were obtained from the Harvard PRIM-H project, who obtained it from Dr. John Woodhouse, Dept. of Geophysics, Harvard University.

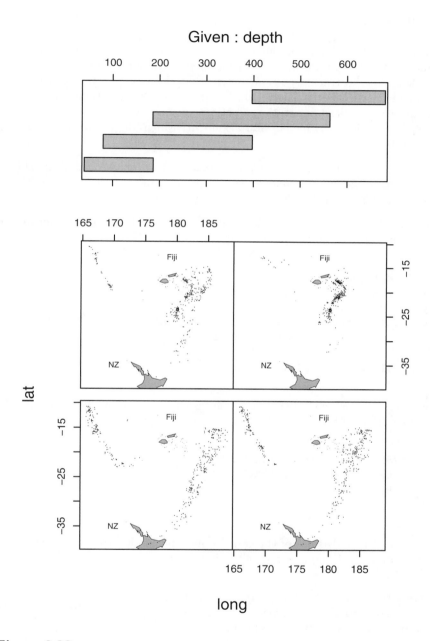

Figure 3.28
A panel function example. An example of using a panel function to add customized
output to each element of a multi-panel plot. A panel function is defined that adds
maps of Fiji and New Zealand to each panel.

3D plots

It is possible to annotate a plot produced using the `persp()` function, but it is more difficult than for most other high-level functions. The important step is to acquire the transformation matrix that the `persp()` function returns. This can be used to transform 3D locations into 2D locations that can be given to the standard annotation functions such as `lines()` and `text()`. The `persp()` function also has an **add** argument, which allows multiple `persp()` plots to be over-plotted.

The following code demonstrates annotation of `persp()` output to add an indication of the summit and access roads to a plot of the Maunga Whau volcano in Auckland New Zealand (see Figure 3.29).*

The first step is to draw the volcano itself and record the 3D transformation matrix in the variable **trans**.

```
> z <- 2 * volcano
> x <- 10 * (1:nrow(z))
> y <- 10 * (1:ncol(z))
> trans <- persp(x, y, z, theta = 135, phi = 30,
                 scale = FALSE, ltheta = -120,
                 box = FALSE)
> box(col="grey", lwd=1)
```

Now a function is defined that uses the transformation matrix to convert 3D locations into 2D locations relative to the existing plot.

```
> trans3d <- function(x,y,z,pmat) {
    tmat <- cbind(x,y,z,1)%*% pmat
    tmat[,1:2] / tmat[,4]
  }
```

The next code makes use of the transformation function to draw a dot at the summit of the volcano and a text label above that.

```
> summit <- trans3d(x[20], y[31], max(z), trans)
> points(summit[1], summit[2], pch=16)
> summitlabel <- trans3d(x[20], y[31], max(z) + 50, trans)
> text(summitlabel[1], summitlabel[2], "Summit")
```

---

*The data are from the `volcano` data set (see page 35) and from the `volcano.accessRoad` `volcano.upDownRoad` `volcano.summitRoad` data sets from the `RGraphics` package.

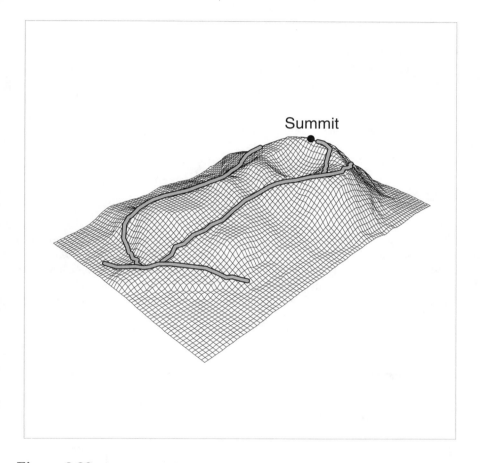

Figure 3.29
Annotating a 3D surface created by `persp()`. Extra points, text, and lines are added
to the 3D plot using the transformation matrix returned by the `persp()` function.

Finally, the transformation function is also used to draw lines representing the roads that provide access to the summit of the volcano.

```
> drawRoad <- function(x, y, z, trans) {
    road <- trans3d(x, y, z, trans)
    lines(road[,1], road[,2], lwd=5)
    lines(road[,1], road[,2], lwd=3, col="grey")
  }
> with(volcano.summitRoad,
       drawRoad(srx, sry, srz, trans))
> with(volcano.upDownRoad,
       {
          clipudx <- udx
          clipudx[udx < 230 & udy < 300 |
                  udx < 150 & udy > 300] <- NA
          drawRoad(clipudx, udy, udz, trans)
       })
> with(volcano.accessRoad,
       drawRoad(arx, ary, arz, trans))
```

This example does demonstrate one of the limitations for annotating `persp()` output, namely that there is no support for automatically hiding output that should not be seen. For example, the drawing of the `upDownRoad` has been manually clipped (see the lines involving the variable `clipudx`) in order to avoid drawing the part of the road that should be obscured because it is behind the main cone of the volcano.

3.5   Creating new plots

There are cases where no existing plot provides a sensible starting point for creating the final plot that the user requires. This section describes how to construct a new plot entirely from scratch for such cases.

The `plot.new()` function is the most basic starting point for producing a traditional graphics plot (the `frame()` function is equivalent). This function sets up the various plotting regions described in Section 3.1.1 and sets both the x-scale and y-scale to $(0, 1)$.* The regions that are set up depend on the

---

*The actual scale setup depends on the current settings for `xaxs` and `yaxs`. With the default settings, the scales are $(-0.04, 1.04)$.

current graphics state settings.

The `plot.window()` function resets the scales in the user coordinate system, given x- and y-ranges via the arguments `xlim` and `ylim`, and the `plot.xy()` function draws data symbols and lines between locations within the plot region.

3.5.1 A simple plot from scratch

In order to demonstrate the use of these functions, the following code produces the simple scatterplot in Figure 1.1 from scratch.

```
> plot.new()
> plot.window(range(pressure$temperature),
            range(pressure$pressure))
> plot.xy(pressure, type="p")
> box()
> axis(1)
> axis(2)
```

The output could be produced by the simple expression `plot(pressure)`, but it shows that the steps in building a plot are available as separate functions as well, which allows the user to have fine control over the construction of a plot.

3.5.2 A more complex plot from scratch

This section describes a slightly more complex example of creating a plot from scratch. The final goal is represented in Figure 3.30 and the steps involved are described below.

This first bit of code generates some data to plot.

```
> groups <- c("cows", "sheep", "horses",
            "elephants", "giraffes")
> males <- sample(1:10, 5)
> females <- sample(1:10, 5)
```

There are several ways that the plot could be created. For this example, it will be created as a single plot. The labels to the left of the plot will be drawn in the margins of the plot, but everything else will be drawn inside the plot region. This next bit of code sets up the figure margins so that there is enough

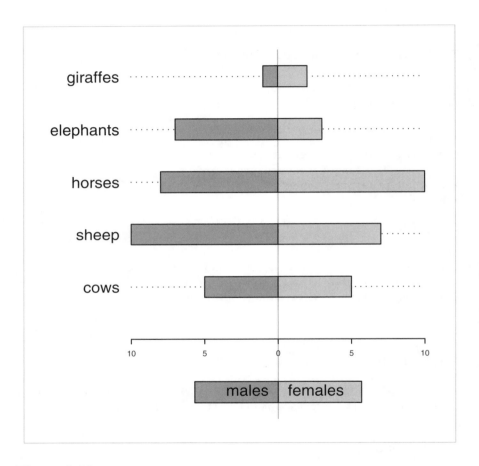

Figure 3.30
A back-to-back barplot from scratch. This demonstrates the use of lower-level plotting functions to produce a novel plot that cannot be produced by an existing high-level function.

room for the labels in the left margin, but all other margins are nice and small (to avoid lots of empty space around the plot).

```
> par(mar=c(0.5, 5, 0.5, 1))
```

Inside the plot region there are seven different rows of output to draw: the five main pairs of bars, the x-axis, and the legend at the bottom. The axis will be drawn at a y-location of 0, the main bars at the y-locations 1:5, and the legend at -1. The following code starts the plot and sets up the appropriate y-scale and x-scale.

```
> plot.new()
> plot.window(xlim=c(-10, 10), ylim=c(-1.5, 5.5))
```

This next bit of code assigns some useful values to variables, including the x-locations of tick-marks on the x-axis, the y-locations of the main bars, and a value representing half the height of the bars.

```
> ticks <- seq(-10, 10, 5)
> y <- 1:5
> h <- 0.2
```

Now some drawing can occur. This next code draws the main part of the plot. Everything is drawn using calls to the low-level functions such as lines(), segments(), mtext(), and axis(). In particular, the main bars are just rectangles produced using rect(). Notice that the x-axis is drawn within the plot region (pos=0).

```
> lines(rep(0, 2), c(-1.5, 5.5), col="grey")
> segments(-10, y, 10, y, lty="dotted")
> rect(-males, y-h, 0, y+h, col="dark grey")
> rect(0, y-h, females, y+h, col="light grey")
> mtext(groups, at=y, adj=1, side=2, las=2)
> par(cex.axis=0.5, mex=0.5)
> axis(1, at=ticks, labels=abs(ticks), pos=0)
```

The final step is to produce the legend at the bottom of the plot. Again, this is just a series of calls to low-level functions, although the bars are sized using strwidth() to ensure that they contain the labels.

```
> tw <- 1.5*strwidth("females")
> rect(-tw, -1-h, 0, -1+h, col="dark grey")
> rect(0, -1-h, tw, -1+h, col="light grey")
> text(0, -1, "males", pos=2)
> text(0, -1, "females", pos=4)
```

This example is particularly customized to the data set involved. It could be made much more general by replacing some constants with variable values (e.g., instead of using 5 because there are five groups in the data set, the code could have a variable `numGroups`). If more than one such plot needs to be made, it makes good sense to also wrap the code within a function. That task is discussed in the next section.

3.5.3   Writing traditional graphics functions

Having made the effort to construct a plot from scratch, it is usually worthwhile encapsulating the calls within a new function and possibly even making it available for others to use. This section briefly describes some of the things to consider when creating a new graphics function built on the traditional graphics system functions.

There are many advantages to developing new graphics functions in the grid graphics system (see Part II) rather than using traditional graphics. Chapter 7 contains a more complete discussion of the issues involved in developing new graphics functions.

Helper functions

There are some helper functions that do no drawing, but are used by the predefined high-level plots to do some of the work in setting up a plot.

The `xy.coords()` function is useful for allowing `x` and `y` arguments to your new function to be flexibly specified (just like the `plot()` function where `y` can be left unspecified and `x` can be a `data.frame`, and so on). This function takes `x` and `y` arguments and creates a standard object containing x-value, y-values, and sensible labels for the axes. There is also an `xyz.coords()` function.

If your plotting function generates multiple sub-plots, the `n2mfrow()` function may be helpful to generate a sensible number of rows and columns of plots, based on the total number of plots to fit on a page.

Another set of useful helper functions are those that calculate values to plot from the raw data (but do no actual drawing). Examples of these sorts of functions are: `boxplot.stats()` used by `boxplot()` to generate five-number summaries; `contourLines()` used by `contour()` to generate contour lines; `nclass.Sturges()`, `nclass.scott()`, and `nclass.FD()` used by `hist()` to generate the number of intervals for a histogram; and `co.intervals()` used by `coplot()` to generate ranges of values for conditioning a data set into panels.

Some high-level functions invisibly return this sort of information too. For example, `boxplot()` returns the combined results from `boxplot.stats()` for all of the boxplots that it produces and `hist()` returns information on the intervals that it creates including the number of data values in each interval.

Argument lists

A common technique when writing a traditional graphics function is to provide an ellipsis argument (. . .) instead of individual graphics state arguments (such as `col` and `lty`). This allows users to specify any state settings (e.g., `col="red"` and `lty="dashed"`) and the new function can pass them straight on to the traditional graphics functions that the new function calls. This avoids having to specify all individual state settings as arguments to the new function. Some care must be taken with this technique because sometimes different graphics functions interpret the same graphics state setting in different ways (the `col` setting is a good example; see Section 3.2). In such cases, it becomes necessary to name the individual graphics state setting as an argument and explicitly pass it on only to other graphics calls that will accept it and respond to it in the desired manner.

Sometimes it is useful for a graphics function to deliberately override the current graphics state settings. For example, a new plot may want to force the `xpd` setting to be `NA` in order to draw lines and text outside of the plot region. In such cases, it is polite for the graphics function to revert the graphics state settings at the end of the function so that users do not get a nasty surprise! A standard technique is to put the following expressions at the start of the new function to restore the graphics state to the settings that existed before the function was called.

```
opar <- par(no.readonly=TRUE)
on.exit(par(opar))
```

Care should be taken to ensure that a new graphics function takes notice of appropriate graphics state settings (e.g., `ann`). This can be a little complicated to implement because it is necessary to be aware of the possibility that the user might specify a setting in the call to the function and that such a setting should override the main graphics state setting. The standard approach is to name the state setting explicitly as an argument to the graphics function and provide the permanent state setting as a default value. See the new graphics function template below for an example of this technique using the `ann` argument. An additional complication is that now there is a state setting that will not be part of the ... argument, so the state setting must be explicitly passed on to any other functions that might make use of it.

Another good technique is to provide arguments that users are used to seeing in other graphics functions — the `main`, `sub`, `xlim`, and `ylim` arguments are good examples of this sort of thing — and a new graphics function should be able to handle missing and non-finite values. The functions `is.na()`, `is.finite()`, and `na.omit()` may be useful for this purpose.

Plot methods

If a new function is for use with a particular type of data, then it is convenient for users if the function is provided as a method for the generic `plot()` function. This allows users to simply call the new function by calling `plot(x)`, where `x` is an object of the relevant class.

A graphics function template

The code in Figure 3.31 is a simple shell that combines some of the basic guidelines from this section. This is just a simplified version of the default `plot()` method. It is far from complete and will not gracefully accept all possible inputs (especially via the ... argument), but it could be used as the starting template for writing a new traditional graphics function.

```
 1 plot.newclass <-
 2   function(x, y=NULL,
 3             main="", sub="",
 4             xlim=NULL, ylim=NULL,
 5             axes=TRUE, ann=par("ann"),
 6             col=par("col"),
 7             ...) {
 8     xy <- xy.coords(x, y)
 9     if (is.null(xlim))
10       xlim <- range(xy$x[is.finite(xy$x)])
11     if (is.null(ylim))
12       ylim <- range(xy$y[is.finite(xy$y)])
13     opar <- par(no.readonly=TRUE)
14     on.exit(par(opar))
15     plot.new()
16     plot.window(xlim, ylim, ...)
17     points(xy$x, xy$y, col=col, ...)
18     if (axes) {
19       axis(1)
20       axis(2)
21       box()
22     }
23     if (ann)
24       title(main=main, sub=sub,
25             xlab=xy$xlab, ylab=xy$ylab, ...)
26 }
```

Figure 3.31

A graphics function template. This code provides a starting point for producing a new graphics function for others to use.

Chapter summary

High-level traditional graphics functions produce complete plots and low-level traditional graphics functions add output to existing plots. There are low-level functions for producing simple output such as lines, rectangles, text, and polygons and also functions for producing more complex output such as axes and legends.

The traditional graphics system creates several regions for drawing the various components of a plot: a plot region for drawing data symbols and lines, figure margins for axes and labels, and so on. Each low-level graphics function produces output in a particular drawing region and most work in the plot region.

There is a traditional graphics system state that consists of settings to control the appearance of output and the arrangement of the drawing regions. There are settings for controlling color, fonts, line styles, data symbol style, and the style of axes. There are several mechanisms for arranging multiple plots on a single page.

It is straightforward to create a complete plot using only low-level graphics functions. This makes it possible to produce a completely new type of plot. It is also possible for the user to define an entirely new graphics function.

Part II

GRID GRAPHICS

4

Trellis Graphics: the Lattice Package

Chapter preview

This chapter describes how to produce Trellis plots using R. There is a description of what Trellis plots are as well as a description of the functions used to produce them. Trellis plots are designed to be easy to interpret and at the same time provide some modern and sophisticated plotting styles, such as multipanel conditioning.

The grid graphics system provides no high-level plotting functions itself, so this chapter also describes the best way to produce a complete plot using the grid system. There are several advantages to producing a plot using the grid system, including greater flexibility in adding further output to the plot, and the ability to interactively edit the plot.

This chapter describes the lattice package, developed by Deepayan Sarkar[54]. Lattice is based on the grid graphics system, but can be used as a complete graphics system in itself and a great deal can be achieved without encountering any of the underlying grid concepts.* This chapter deals with lattice as a self-contained system consisting of functions for producing complete plots, functions for controlling the appearance of the plots, and functions for opening and closing devices. Section 5.8 and Section 6.7 describe some of the benefits that can be gained from viewing lattice plots as grid output and dealing directly with the grid concepts and objects that underly the lattice system.

---

*To give Deepayan proper credit, lattice uses grid only to *render* plots. Lattice performs a lot of work itself to deconstruct formulae, rearrange the data, and manage many user-settable options.

The graphics functions that make up the lattice graphics system are provided in an add-on package called `lattice`. The lattice system is loaded into R as follows.

```
> library(lattice)
```

The lattice package implements the Trellis Graphics system[6] with some novel extensions. The Trellis Graphics system has a large number of sophisticated features and many of these are described in this section, but more information, examples, and background are available from the Trellis Display web site:

`http://cm.bell-labs.com/cm/ms/departments/sia/project/trellis/index.html`

4.1   The lattice graphics model

In simple usage, lattice functions appear to work just like traditional graphics functions where the user calls a function and output is generated on the current device. The following example plots the locations of 1000 earthquakes that have occurred in the Pacific Ocean (near Fiji) since 1964 (see Figure 4.1).*

```
> xyplot(lat ~ long, data=quakes, pch=".")
```

It is perfectly valid to use lattice this way; however, lattice graphics functions do not produce graphical output directly. Instead they produce an object of class `"trellis"`, which contains a description of the plot. The `print()` method for objects of this class does the actual drawing of the plot. This can be demonstrated quite easily. For example, the following code creates a `trellis` object, but does not draw anything.

```
> tplot <- xyplot(lat ~ long, data=quakes, pch=".")
```

The result of the call to `xyplot()` is assigned to the variable `tplot` so it is not printed. The plot can be drawn by calling print on the `trellis` object (the result is exactly the same as Figure 4.1).

```
> print(tplot)
```

---

*The data are available as the data set `quakes` in the `datasets` package.

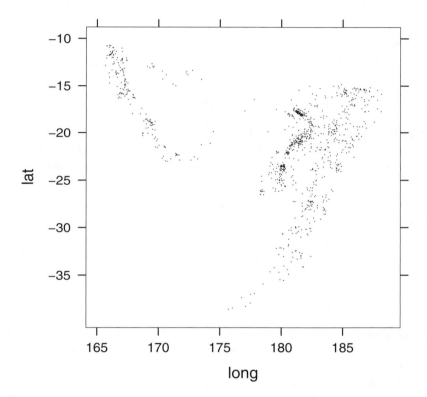

Figure 4.1
A scatterplot using lattice (showing the locations of earthquakes in the Pacific Ocean). A basic lattice plot has a very similar appearance to an analogous traditional plot.

This design makes it possible to work with the `trellis` object and modify it using the `update()` method for `trellis` objects, which is an alternative to modifying the original R expression used to create the `trellis` object. The following code demonstrates this idea by modifying the `trellis` object `tplot` to redefine the main title of the plot (it was empty). The resulting output is shown in Figure 4.2. A subtle change to look for is the fact that extra space has been introduced to allow room for adding the new main title text (the height of the plot region is slightly smaller compared to Figure 4.1).

```
> update(tplot,
        main="Earthquakes in the Pacific Ocean\n(since 1964)")
```

The side-effect of the code above is to produce new output that is a modification of the original plot, represented by `tplot`. However, it is important to remember that `tplot` has not been changed in any way (typing `tplot` again will produce output like Figure 4.1 again). In order to retain an R object representing the modified plot, the user must assign the value returned by the `update()` function, as in the following code.

```
> tplot2 <-
    update(tplot,
        main="Earthquakes in the Pacific Ocean (since 1964)")
```

4.1.1 Lattice devices

For each graphics device, lattice maintains its own set of graphical parameter settings that control the appearance of plots (colors of lines, fonts for text, and many more — see Section 4.3)*. The default settings depend on the type of device being opened (e.g., the settings are different for a PostScript device compared to a PDF device). In simple usage this causes no problems, because lattice automatically initializes these settings the first time that lattice output is produced on a device. If it is necessary to control the initial values for these settings the `trellis.device()` function can be used to explicitly open a device with specific lattice graphical parameter settings (or just to enforce specific lattice settings on an existing device). Section 4.3 describes more functions for manipulating the lattice graphical parameter settings.

---

*One of the features of Trellis Graphics is that carefully selected default settings are provided for colors, data symbols, and so on. These settings are selected to maximize the interpretability of plots and are based on principles of human perception[15].

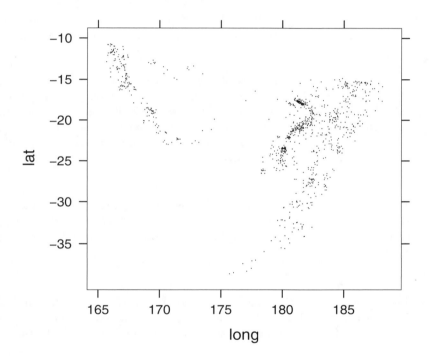

Figure 4.2
The result of modifying a lattice object. Lattice creates an object representing the plot. If this object is modified, the plot is redrawn. This figure shows the result of modifying the object representing the plot in Figure 4.1 to add a title to the plot.

4.2 Lattice plot types

Lattice provides functions to produce a number of standard plot types, plus some more modern and specialized plots. Table 4.1 describes the functions that are available and Figure 4.3 provides a basic idea of the sort of output that they produce.

There are a number of functions that produce output very similar to the output of functions in the traditional graphics system, but there are three possible reasons for using lattice functions instead of the traditional counterparts:

1. The default appearance of the lattice plots is superior in some areas. For example, the default colors and the default data symbols have been deliberately chosen to make it easy to distinguish between groups when more than one data series is plotted. There are also some subtle things such as the fact that tick labels on the y-axes are written horizontally by default, which makes them easier to read.

2. The lattice plot functions can be extended in several very powerful ways. For example, several data series can be plotted at once in a convenient manner and multiple panels of plots can be produced easily (see Section 4.2.1).

3. The output from lattice functions is grid output, so many powerful grid features are available for annotating, editing, and saving the graphics output. See Section 5.8 and Section 6.7 for examples of these features.

Most of the lattice plotting functions provide a very long list of arguments and produce a wide range of different types of output. Many of the arguments are shared by different functions and the on-line help for the `xyplot()` function provides an explanation of these standard arguments. The following sections address some of the important shared arguments, but for a full explanation of all arguments, the documentation for each specific function should be consulted. The next section discusses two important arguments, `formula` and `data`. The use of several other arguments is demonstrated in Section 4.2.2 in the context of a more complex example. Section 4.3 mentions the `par.settings` argument and Section 4.4 describes the `layout` argument. Section 4.5 describes the `panel` and `strip` arguments.

Table 4.1

The plotting functions available in lattice

Lattice Function	Description	Traditional Analogue
barchart()	Barcharts	barplot()
bwplot()	Boxplots Box-and-whisker plots	boxplot()
densityplot()	Conditional kernel density plots Smoothed density estimate	*none*
dotplot()	Dotplots Continuous versus categorical	dotchart()
histogram()	Histograms	hist()
qqmath()	Quantile–quantile plots Data set versus theoretical distribution	qqnorm()
stripplot()	Stripplots One-dimensional scatterplot	stripchart()
qq()	Quantile–quantile plots Data set versus data set	qqplot()
xyplot()	Scatterplots	plot()
levelplot()	Level plots	image()
contourplot()	Contour plots	contour()
cloud()	3-dimensional scatterplot	*none*
wireframe()	3-dimensional surfaces	persp()
splom()	Scatterplot matrices	pairs()
parallel()	Parallel coordinate plots	*none*

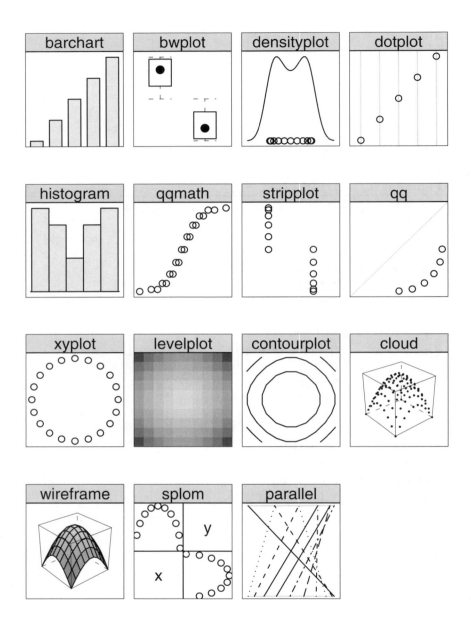

Figure 4.3
Plot types available in lattice. The name of the function used to produce the different plot types is shown in the strip above each plot.

4.2.1   The `formula` argument and multipanel conditioning

In most cases, the first argument to the lattice plotting functions is an R formula (see Section A.2.6) that describes which variables to plot. The simplest case has already been demonstrated. A formula of the form `y ~ x` plots variable `y` against variable `x`. There are some variations for plots of only one variable or plots of more than two variables. For example, for the `bwplot()` function, the formula can be of the form `~ x` and for the `cloud()` and `wireframe()` functions something of the form `z ~ x * y` is required to specify the three variables to plot. Another useful variation is the ability to specify multiple y-variables. Something of the form `y1 + y2 ~ x` produces a plot of both the `y1` variable and the `y2` variable against `x`. Multiple x-variables can be specified as well.

The second argument to a lattice plotting function is typically `data`, which allows the user to specify a data frame within which lattice can find the variables specified in the formula.

One of the very powerful features of Trellis Graphics is the ability to specify conditioning variables within the formula argument. Something of the form `y ~ x | g` indicates that several plots should be generated, showing the variable `y` against the variable `x` for each level of the variable `g`. In order to demonstrate this feature, the following code produces several scatterplots, with each scatterplot showing the locations of earthquakes that occurred within a particular depth range (see Figure 4.4). First of all, a new variable `depthgroup` is defined, which is a binning of the original `depth` variable in the `quakes` data set.

```
> depthgroup <- equal.count(quakes$depth, number=3, overlap=0)
```

Now this `depthgroup` variable can be used to produce a scatterplot for each depth range.

```
> xyplot(lat ~ long | depthgroup, data=quakes, pch=".")
```

In the Trellis terminology, the plot in Figure 4.4 consists of three *panels*. Each panel in this case contains a scatterplot and above each panel there is a *strip* that presents the level of the conditioning variable.

There can be more than one conditioning variable in the formula argument, in which case a panel is produced for each combination of the conditioning variables. An example of this is given in Section 4.2.2.

The most natural type of variable to use as a conditioning variable is a categorical variable (factor), but there is also support for using a continuous

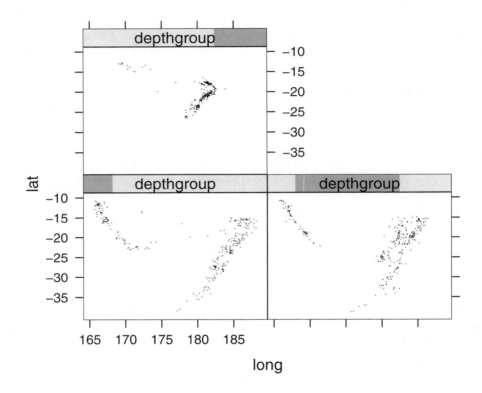

Figure 4.4
A lattice multipanel conditioning plot. A single function call produces several scatterplots of the locations of earthquakes for different earthquake depths.

(numeric) conditioning variable. For this purpose, Trellis Graphics introduces the concept of a *shingle*. This is a continuous variable with a number of ranges associated with it. The ranges are used to split the continuous values into (possibly overlapping) groups. The `shingle()` function can be used to explicitly control the ranges, or the `equal.count()` function can be used to generate ranges automatically given a number of groups (as was done to produce the `depthgroup` variable above).

4.2.2 A nontrivial example

This section describes an example that makes use of some of the common arguments to the lattice plotting functions to produce a more complex final result (see Figure 4.5). First of all, another grouping variable, `magnitude`, is defined, which is a shingle indicating whether an earthquake is big or small.

```
> magnitude <- equal.count(quakes$mag, number=2, overlap=0)
```

The plot is still produced from a single function call, but there are two conditioning variables, so there is a panel for each possible combination of depth and magnitude. A title and axis labels have been specified for the plot using the `main`, `xlab`, and `ylab` arguments. The `between` argument has been used to introduce a vertical gap between the top row of panels (big earthquakes) and the bottom row of panels (small earthquakes). The `par.strip.text` argument is used to control the size of text in the strips above each panel. The `scales` argument is used to control the drawing of axis labels; in this case the specification says that the x-axis labels should go at the bottom for both panels. This is to avoid the axis tick marks interfering with the main title. Finally, the `par.settings` argument is used to control the size of the tick labels on the axes.

```
> xyplot(lat ~ long | depthgroup * magnitude,
        data=quakes,
        main="Fiji Earthquakes",
        ylab="latitude", xlab="longitude",
        pch=".",
        scales=list(x=list(alternating=c(1, 1, 1))),
        between=list(y=1),
        par.strip.text=list(cex=0.7),
        par.settings=list(axis.text=list(cex=0.7)))
```

This example demonstrates that it is possible to have very fine control over many aspects of a lattice plot, given sufficient willingness to learn about all of the arguments that are available.

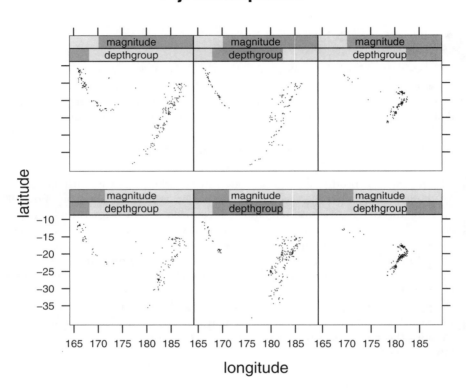

Figure 4.5
A complex lattice plot. There are a large number of arguments to lattice plotting
functions to allow control over many details of a plot, such as the text to use for
labels and titles, the size and placement of axis tick labels, and the size of the gaps
between columns and rows of panels.

4.3 Controlling the appearance of lattice plots

An important feature of Trellis Graphics is the careful selection of default settings that are provided for many of the features of lattice plots. For example, the default data symbols and colors used to distinguish between different data series have been chosen so that it is easy to visually discriminate between them. Nevertheless, it is still sometimes desirable to be able to make alterations to the default settings for aspects like color and text size. It is also useful to be able to control the layout or arrangement of the components (panels and strips) of a lattice plot, but that is dealt with separately in Section 4.4. This section is only concerned with graphical parameters that control colors, line types, fonts and the like.

The lattice graphical parameter settings consist of a large list of parameter groups and each parameter group is a list of parameter settings. For example, there is a `plot.line` parameter group consisting of `col`, `lty`, and `lwd` settings to control the color, line type, and line width for lines drawn between data locations. There is a separate `plot.symbol` group consisting of `cex`, `col`, `font`, and `pch` settings to control the size, shape, and color of data symbols. The settings in each parameter group affect some aspect of a lattice plot: some have a "global" effect; for example, the `fontsize` settings affect all text in a plot; some are more specific; for example, the `strip.background` setting affects the background color of strips; and some only affect a certain aspect of a certain sort of plot; for example, the `box.dot` settings affect only the dot that is plotted at the median value in boxplots.

A separate list of graphical parameters is maintained for each graphics device. Changes to parameter settings (see below) only affect the current device.

The function `show.settings()` produces a picture representing some of the current graphical parameter settings. Figure 4.6 shows the settings for a black-and-white PostScript device.

The current value of graphical parameter settings can be obtained using the `trellis.par.get()` function. For a list of all current graphical parameter settings, type `trellis.par.get()`. If a name is specified as the argument to this function, then only the relevant settings are returned. The following code shows how to obtain only the `fontsize` group of settings (the output is on page 139).

```
> trellis.par.get("fontsize")
```

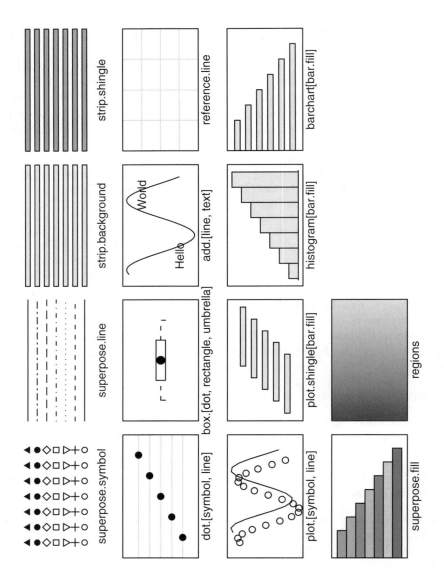

Figure 4.6
Some default lattice settings for a black-and-white PostScript device. This figure
was produced by the lattice function show.settings().

```
$text
[1] 9

$points
[1] 8
```

There are two ways to set new values for graphical parameters. The values can be set persistently (i.e., they will affect all subsequent plots until a new setting is specified) using the `trellis.par.set()` function, or they can be set temporarily for a single plot by specifying settings as an argument to a plotting function.

The `trellis.par.set()` function can be used in several ways. For back-compatibility with the original implementation of Trellis, it is possible to provide a name as the first argument and a list of settings as the second argument. This will modify the values for one parameter group.

A new approach is to provide a list of lists that can be used to modify multiple parameter groups at once. Lattice also introduces the concept of *themes*, which is a comprehensive and coherent set of graphical parameter values. It is possible to specify such a theme and enforce a new "look and feel" for a plot in one function call. Lattice currently provides one such theme via the `col.whitebg()` function. It is also possible to obtain the default theme for a particular device using the `canonical.theme()` function.

The following code demonstrates how to use `trellis.par.set()` in either the backwards-compatible, one-parameter-group-at-a-time way, or the new list-of-lists way, to specify `fontsize` settings.

```
> trellis.par.set("fontsize", list(text=14, points=10))
> trellis.par.set(list(fontsize=list(text=14, points=10)))
```

The theme approach is usually more convenient, especially when setting only one value within a parameter group. For example, the following code demonstrates the difference between the two approaches for modifying just the `text` setting within the `fontsize` parameter group (old way first, new way second).

```
> fontsize <- trellis.par.get("fontsize")
> fontsize$text <- 20
> trellis.par.set("fontsize", fontsize)

> trellis.par.set(list(fontsize=list(text=20)))
```

The concept of themes is an example of a lattice-specific extension to the original Trellis Graphics system.

The other way to modify lattice graphical parameter settings is on a per-plot basis, by specifying a `par.settings` argument in the call to a plotting function. The value for this argument should be a theme (a list of lists). Such a setting will only be enforced for the relevant plot and will not affect any subsequent plots. The following code demonstrates how to modify the `fontsize` settings just for a single plot.

```
> xyplot(lat ~ long, data=quakes,
         par.settings=list(fontsize=list(text=14, points=10)))
```

4.4   Arranging lattice plots

There are two types of arrangements to consider when dealing with lattice plots: the arrangement of panels and strips within a single lattice plot; and the arrangement of several complete lattice plots together on a single page.

In the first case (the arrangement of panels and strips within a single plot) there are two useful arguments that can be specified in a call to a lattice plotting function: the `layout` argument and the `aspect` argument.

The `layout` argument consists of up to three values. The first two indicate the number of columns and rows of panels on each page and the third value indicates the number of pages. It is not necessary to specify all three values, as lattice provides sensible default values for any unspecified values. The following code produces a variation on Figure 4.4 by explicitly specifying that there should be a single column of three panels via the `layout` argument, and that each panel must be "square" via the `aspect` argument. The `index.cond` argument has also been used to specify that the panels should be ordered from top to bottom (see Figure 4.7).

```
> xyplot(lat ~ long | depthgroup, data=quakes, pch=".",
         layout=c(1, 3), aspect=1, index.cond=list(3:1))
```

The `aspect` argument specifies the aspect ratio (height divided by width) for the panels. The default value is `"fill"`, which means that panels expand to occupy as much space as possible. In the example above, the panels were all forced to be square by specifying `aspect=1`. This argument will also accept the special value `"xy"`, which means that the aspect ratio is calculated to satisfy the "banking to 45 degrees" rule proposed by Bill Cleveland[13].

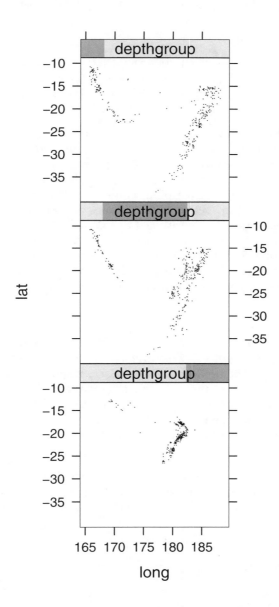

Figure 4.7
Controlling the layout of lattice panels. Lattice arranges panels in a sensible way by default, but there are several ways to force the panels to be arranged in a particular layout. This figure shows a custom arrangement of the panels in the plot from Figure 4.4.

As with the choice of colors and data symbols, a lot of work is done to select sensible default values for the arrangement of panels, so in many cases nothing special needs to be specified.

Another issue in the arrangement of a single lattice plot is the placement and structure of the key or legend. This can be controlled using the `auto.key` or `key` argument to plotting functions, which will accept complex specifications of the contents, layout, and positioning of the key.

The problem of arranging multiple lattice plots on a page requires a different approach. A `trellis` object must be created (but not plotted) for each lattice plot, then the `print()` function is called, supplying arguments to specify the position of each plot. The following code provides a simple demonstration using the average yearly number of sunspots from 1749 to 1983, available as the `sunspots` data set in the `datasets` package (see Figure 4.8). Two lattice plots are produced and then positioned one above the other on a page. The `position` argument is used to specify their location, (`left`, `bottom`, `right`, `top`), as a proportion of the total page, and the `more` argument is used in the first `print()` call to ensure that the second `print()` call draws on the same page. The `scales` argument is also used to draw the x-axis at the top of the top plot.

```
> spots <- by(sunspots, gl(235, 12, lab=1749:1983), mean)
> plot1 <- xyplot(spots ~ 1749:1983, xlab="", type="l",
                  main="Average Yearly Sunspots",
                  scales=list(x=list(alternating=2)))
> plot2 <- xyplot(spots ~ 1749:1983, xlab="Year", type="l")
> print(plot1, position=c(0, 0.2, 1, 1), more=TRUE)
> print(plot2, position=c(0, 0, 1, 0.33))
```

Section 5.8 describes additional options for controlling the arrangements of panels within a lattice plot, and more flexible options for arranging multiple lattice plots, using the concepts and facilities of the grid system.

4.5   Annotating lattice plots

In the original Trellis Graphics system, plots are completely self-contained. There is no real concept of adding output to a plot once the plot has been drawn. This constraint has been lifted in lattice, though the traditional approach is still supported.

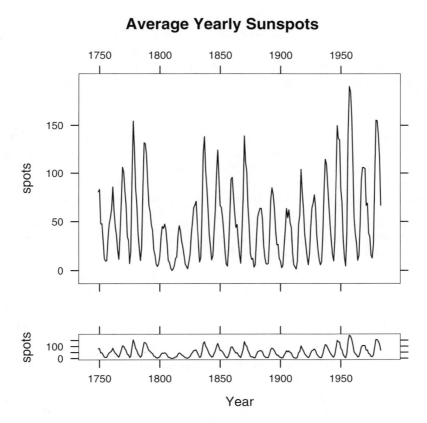

Figure 4.8
Arranging multiple lattice plots. This shows two separate lattice plots arranged together on a single page.

4.5.1   Panel functions and strip functions

The `trellis` object that is produced by a lattice plotting function is a complete description of a plot. The usual way to add extra output to a plot (e.g., add text labels to data symbols), is to add extra information to the `trellis` object. This is achieved by specifying a *panel function* via the `panel` argument of lattice plotting functions.

The panel function is called for each panel in a lattice plot. All lattice plotting functions have a default panel function, which is usually the name of the function with a "`panel.`" prefix. For example, the default panel function for the `xyplot()` function is `panel.xyplot()`. The default panel function draws the default contents for a panel so it is typical to call this default as part of a custom panel function.

The arguments available to the panel function differ depending on the plotting function. The documentation for individual panel functions should be consulted for full details, but some common arguments to expect are `x` and `y` (and possibly `z`), giving locations at which to plot data symbols, and `subscripts`, which provides the indices used to obtain the subset of the data for each panel.

In addition to the panel function, it is possible to specify a *prepanel function* for controlling the scaling and size of panels and a *strip function* for controlling what gets drawn in the strips of a lattice plot.

The following code provides a simple demonstration of the use of panel, prepanel and strip functions. The plot is a lattice multi-panel scatterplot with text labels added to the data points and a custom strip showing both levels of the conditioning variable with the relevant level bold and the other level grey (see Figure 4.9).

The panel function calls the default `panel.xyplot()` to draw data symbols, then calls `ltext()` to draw the labels. Because lattice is based on grid, traditional graphics functions will not work in a panel function (though see Appendix B for a way around this constraint). However, there are several lattice functions that correspond to traditional functions and can be used in much the same way as the corresponding traditional functions. The names of the lattice analogues are the traditional function names with an "`l`" prefix added. In this case, the code draws letters as the labels, using the `subscripts` argument to select an appropriate subset. The labels are drawn slightly to the left of and above the data symbols by subtracting 1 from the `x` values and adding 1 to the `y` values.

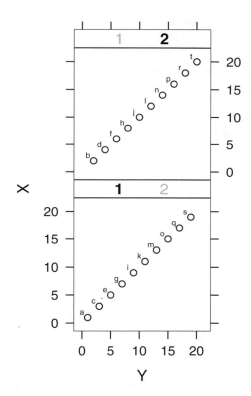

Figure 4.9
Annotating a lattice plot using panel and strip functions. The text labels have been
added beside the data symbols using a custom panel function and the bold and grey
numerals in the strips have been produced using a custom strip function.

```
> myPanel <- function(x, y, subscripts, ...) {
    panel.xyplot(x, y, ...)
    ltext(x - 1, y + 1, letters[subscripts], cex=0.5)
  }
```

The strip function also uses `ltext()`. Locations within the strip are based on a "normalized" coordinate system with the location (0, 0) at the bottom-left corner and (1, 1) at the top-right corner. The font face and color for the text is calculated using the `which.panel` argument. This supplies the current level for each conditioning variable in the panel.

```
> myStrip <- function(which.panel, ...) {
    font <- rep(1, 2)
    font[which.panel] <- 2
    col=rep("grey", 2)
    col[which.panel] <- "black"
    llines(c(0, 1, 1, 0, 0), c(0, 0, 1, 1, 0))
    ltext(c(0.33, 0.66), rep(0.5, 2), 1:2,
          font=font, col=col)
  }
```

The prepanel function calculates the limits of the scales for each panel by extending the range of data by 1 unit (this allows room for the text labels that are added in the panel function).

```
> myPrePanel <- function(x, y, ...) {
    list(xlim=c(min(x) - 1, max(x) + 1),
         ylim=c(min(y) - 1, max(y) + 1))
  }
```

We now generate some data to plot and create the plot using `xyplot()`, with the special panel functions provided as arguments. The final result is shown in Figure 4.9.

```
> X <- 1:20
> Y <- 1:20
> G <- factor(rep(1:2, 10))

> xyplot(X ~ Y | G, aspect=1, layout=c(1, 2),
         panel=myPanel, strip=myStrip,
         prepanel=myPrePanel)
```

A great deal more can be done with panel functions using grid concepts and functions. See Sections 5.8 and 6.7 for some examples.

4.5.2 Adding output to a lattice plot

Unlike in the original Trellis implementation, it is also possible to add output to a complete lattice plot (i.e., without using a panel function). The function `trellis.focus()` can be used to return to a particular panel or strip of the current lattice plot in order to add further output using, for example, `llines()` or `lpoints()`. The function `trellis.panelArgs()` may be useful for retrieving the arguments (including the data) used to originally draw the panel. Also, the `trellis.identify()` function provides basic mouse interaction for labelling data points within a panel. Again, Sections 5.8 and 6.7 show how grid provides more flexibility for navigating to different parts of a lattice plot and for adding further output.

4.6 Creating new lattice plots

The lattice plotting functions have many arguments and are very flexible in the variety of output that they can produce. However, lattice is not designed to be the best environment for developing new types of graphical display. For example, there is no mechanism for adding new graphical parameters to the list of values that control the appearance of plots (see Section 4.3).

Nevertheless, a lot can be done by defining a panel function that does not just add extra output to the default output, but replaces the default output with some sort of completely different display. For example, the lattice `dotplot()` function is really only a call to the `bwplot()` function with a different panel function supplied.

Users wanting to develop a new lattice plotting function along these lines are advised to read Chapter 5 to gain an understanding of the grid system that is used in the production of lattice output.

Chapter summary

The lattice package implements and extends the Trellis graphics system for producing complete statistical plots. This system provides
most standard plot types and a number of modern plot types with
several important extensions. For a start, the layout and appearance
of the plots is designed to maximize readability and comprehension of
the information represented in the plot. Also, the system provides a
feature called multipanel conditioning, which produces multiple panels
of plots from a single data set, where each panel contains a different
subset of the data. The lattice functions provide an extensive set of
arguments for customizing the detailed appearance of a plot and there
are functions that allow the user to add further output to a plot.

5

The Grid Graphics Model

Chapter preview

This chapter describes the fundamental tools that grid provides for drawing graphical scenes (including plots). There are basic features such as functions for drawing lines, rectangles, and text, together with more sophisticated and powerful concepts such as viewports, layouts, and units, which allow basic output to be located and sized in very flexible ways.

This chapter is useful for drawing a wide variety of pictures, including statistical plots from scratch, and for adding output to lattice plots.

The functions that make up the grid graphics system are provided in an add-on package called `grid`. The grid system is loaded into R as follows.

```
> library(grid)
```

In addition to the standard on-line documentation available via the `help()` function, grid provides both broader and more in-depth on-line documentation in a series of vignettes, which are available via the `vignette()` function.

The grid graphics system only provides low-level graphics functions. There are no high-level functions for producing complete plots. Section 5.1 briefly introduces the concepts underlying the grid system, but this only provides an indication of how to work with grid and some of the things that are possible. An effective direct use of grid functions requires a deeper understanding of the grid system (see later sections of this chapter and Chapter 6).

The lattice package described in Chapter 4 provides a good demonstration of the high-level results that can be achieved using grid. Other examples in this book are Figure 1.7 in Chapter 1 and Figures 7.1 and 7.18 in Chapter 7.

5.1   A brief overview of grid graphics

This chapter describes how to use grid to produce graphical output. There are functions to produce basic output, such as lines and rectangles and text, and there are functions to establish the context for drawing, such as specifying where output should be placed and what colors and fonts to use for drawing.

Like the traditional system, all grid output occurs on the current device,* and later output obscures any earlier output that it overlaps (i.e.,output follows the "painters model"). In this way, images can be constructed incrementally using grid by calling functions in sequence to add more and more output.

There are grid functions to draw primitive graphical output such as lines, text, and polygons, plus some slightly higher-level graphical components such as axes (see Section 5.2). Complex graphical output is produced by making a sequence of calls to these primitive functions.

The colors, line types, fonts, and other aspects that affect the appearance of graphical output are controlled via a set of graphical parameters (see Section 5.4).

Grid provides no predefined regions for graphical output, but there is a powerful facility for defining regions, based on the idea of a *viewport* (see Section 5.5). It is quite simple to produce a set of regions that are convenient for producing a single plot (see the example in the next section), but it is also possible to produce very complex sets of regions such as those used in the production of Trellis plots (see Chapter 4).

All viewports have a large set of coordinate systems associated with them so that it is possible to position and size output in physical terms (e.g., in centimeters) as well as relative to the scales on axes, and in a variety of other ways (see Section 5.3).

All grid output occurs relative to the current viewport (region) on a page. In order to start a new page of output, the user must call the **grid.newpage()**

---

*See Section 1.3.1 for information on devices and selecting a current device when more than one device is open.

function. The function `grid.prompt()` controls whether the user is prompted when moving to a new page.

As well as the side effect of producing graphical output, grid graphics functions produce objects representing output. These objects can be saved to produce a persistent record of a plot, and other grid functions exist to modify these graphical objects (for example, it is possible to interactively edit a plot). It is also possible to work entirely with graphical descriptions, without producing any output. Functions for working with graphical objects are described in detail in Chapter 6.

5.1.1   A simple example

The following example demonstrates the construction of a simple scatterplot using grid. This is more work than a single function call to produce the plot, but it shows some of the advantages that can be gained by producing the plot using grid.

This example uses the **pressure** data to produce a scatterplot much like that in Figure 1.1.

Firstly, some regions are created that will correspond to the "plot region" (the area within which the data symbols will be drawn) and the "margins" (the area used to draw axes and labels).

The following code creates two viewports. The first viewport is a rectangular region that leaves space for 5 lines of text at the bottom, 4 lines of text at the left side, 2 lines at the top, and 2 lines to the right. The second viewport is in the same location as the first, but it has x- and y-scales corresponding to the range of the pressure data to be plotted.

```
> pushViewport(plotViewport(c(5, 4, 2, 2)))
> pushViewport(dataViewport(pressure$temperature,
                            pressure$pressure,
                            name="plotRegion"))
```

The following code draws the scatterplot one piece at a time. Grid output occurs relative to the most recent viewport, which in this case is the viewport with the appropriate axis scales. The data symbols are drawn relative to the x- and y-scales, a rectangle is draw around the entire plot region, and x- and y-axes are drawn to represent the scales.

```
> grid.points(pressure$temperature, pressure$pressure,
              name="dataSymbols")
> grid.rect()
> grid.xaxis()
> grid.yaxis()
```

Adding labels to the axes demonstrates the use of the different coordinate systems available. The label text is drawn outside the edges of the plot region and is positioned in terms of a number of lines of text (i.e.,the height that a line of text would occupy).

```
> grid.text("temperature", y=unit(-3, "lines"))
> grid.text("pressure", x=unit(-3, "lines"), rot=90)
```

The obvious result of running the above code is the graphical output (see the top-left image in Figure 5.1). Less obvious is the fact that several objects have been created. There are objects representing the viewport regions and there are objects representing the graphical output. The following code makes use of this fact to modify the plotting symbol from a circle to a triangle (see the top-right image in Figure 5.1). The object representing the data symbols was named "dataSymbols" (see the code above) and this name is used to find that object and modify it using the grid.edit() function.

```
> grid.edit("dataSymbols", pch=2)
```

The next piece of code makes use of the objects representing the viewports. The upViewport() and downViewport() functions are used to navigate between the different viewport regions to perform some extra annotations. First of all, a call to the upViewport() function is used to go back to working within the entire device so that a dashed rectangle can be drawn around the complete plot. Next, the downViewport() function is used to return to the plot region to add a text annotation that is positioned relative to the scale on the axes of the plot (see bottom-right image in Figure 5.1).

```
> upViewport(2)
> grid.rect(gp=gpar(lty="dashed"))
> downViewport("plotRegion")
> grid.text("Pressure (mm Hg)\nversus\nTemperature (Celsius)",
            x=unit(150, "native"), y=unit(600, "native"))
```

The final scatterplot is still quite simple in this example, but the techniques that were used to produce it are very general and powerful. It is possible to

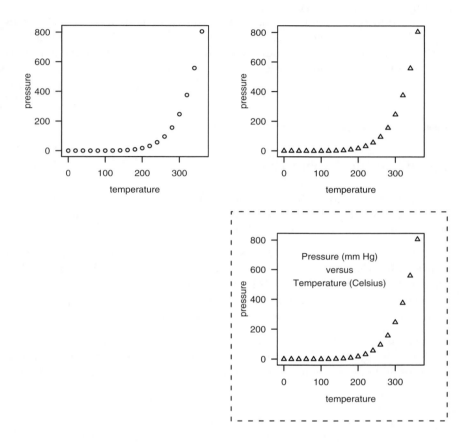

Figure 5.1
A simple scatterplot produced using grid. The top-left plot was constructed from
a series of calls to primitive grid functions that produce graphical output. The
top-right plot shows the result of calling the `grid.edit()` function to interactively
modify the plotting symbol. The bottom-right plot was created by making calls to
`upViewport()` and `downViewport()` to navigate between different drawing regions
and adding further output (a dashed border and text within the plot).

produce a very complex plot, yet still have complete access to modify and add to any part of the plot.

In the remaining sections of this chapter, and in Chapter 6, the basic grid concepts of viewports and units are discussed in full detail. A complete understanding of the grid system will be useful in two ways: it will allow the user to produce very complex images from scratch (the issue of making them available to others is addressed in Chapter 7) and it will allow the user to work effectively with (e.g., modify and add to) complex grid output that is produced by other people's code (e.g. lattice plots).

5.2   Graphical primitives

The most simple grid functions to understand are those that draw something. There are a set of grid functions for producing basic graphical output such as lines, circles, and text.* Table 5.1 lists the full set of these functions.

The first arguments to most of these functions is a set of locations and dimensions for the graphical object to draw. For example, `grid.rect()` has arguments x, y, width, and height for specifying the locations and sizes of the rectangles to draw. An important exception is the `grid.text()` function, which requires the text to draw as its first argument.

In most cases, multiple locations and sizes can be specified and multiple primitives will be produced in response. For example, the following function call produces 100 circles because 100 locations and radii are specified (see Figure 5.2).

```
> grid.circle(x=seq(0.1, 0.9, length=100),
              y=0.5 + 0.4*sin(seq(0, 2*pi, length=100)),
              r=abs(0.1*cos(seq(0, 2*pi, length=100)))))
```

The `grid.move.to()` and `grid.line.to()` functions are unusual in that they both only accept one location. These functions refer to and modify a "current location." The `grid.move.to()` function sets the current location and `grid.line.to()` draws from the current location to a new location, then sets

---

*All of these functions are of the form `grid.*()` and, for each one, there is a corresponding `*Grob()` function that creates an object containing a description of primitive graphical output, but does not draw anything. The `*Grob()` versions are addressed fully in Chapter 6.

Table 5.1

Graphical primitives in grid. This is the complete set of low-level functions that produce graphical output. For each function that produces graphical output (left-most column), there is a corresponding function that returns a graphical object containing a description of graphical output instead of producing graphical output (right-most column). The latter set of functions is described further in Chapter 6.

Function to Produce Output	Description	Function to Produce Object
`grid.move.to()`	Set the current location	`moveToGrob()`
`grid.line.to()`	Draw a line from the current location to a new location and reset the current location.	`lineToGrob()`
`grid.lines()`	Draw a single line through multiple locations in sequence.	`linesGrob()`
`grid.segments()`	Draw multiple lines between pairs of locations.	`segmentsGrob()`
`grid.rect()`	Draw rectangles given locations and sizes.	`rectGrob()`
`grid.circle()`	Draw circles given locations and radii.	`circleGrob()`
`grid.polygon()`	Draw polygons given vertexes.	`polygonGrob()`
`grid.text()`	Draw text given strings, locations and rotations.	`textGrob()`
`grid.arrows()`	Draw arrows at either end of lines given locations or an object describing lines.	`arrowsGrob()`
`grid.points()`	Draw data symbols given locations.	`pointsGrob()`
`grid.xaxis()`	Draw x-axis.	`xaxisGrob()`
`grid.yaxis()`	Draw y-axis.	`yaxisGrob()`

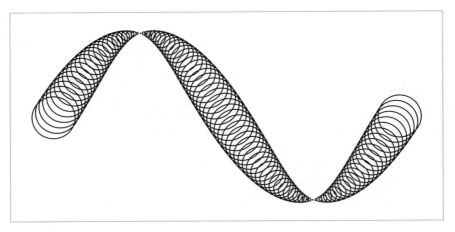

Figure 5.2
Primitive grid output. A demonstration of basic graphical output produced using
a single call to the `grid.circle()` function. There are 100 circles of varying sizes,
each at a different (`x, y`) location.

the current location to be the new location. The current location is not used
by the other drawing functions*. In most cases, `grid.lines()` will be more
convenient, but `grid.move.to()` and `grid.line.to()` are useful for drawing
lines across multiple viewports (an example is given in Section 5.5.1).

The `grid.arrows()` function is used to add arrows to lines. A single line
can be specified by `x` and `y` locations (through which a line will be drawn),
or the `grob` argument can be used to specify an object that describes one or
more lines (produced by `linesGrob()`, `segmentsGrob()`, or `lineToGrob()`).
In the latter case, `grid.arrows()` will add arrows at the ends of the line(s).
The following code demonstrates the different uses (see Figure 5.3). The first
`grid.arrows()` call specifies locations via the `x` and `y` arguments to produce
a single line, at the end of which an arrow is drawn. The second call specifies
a `segments` graphical object via the `grob` argument, which describes three
lines, and an arrow is added to the end of each of these lines.

```
> angle <- seq(0, 2*pi, length=50)
> grid.arrows(x=seq(0.1, 0.5, length=50),
              y=0.5 + 0.3*sin(angle))
> grid.arrows(grob=segmentsGrob(6:8/10, 0.2, 7:9/10, 0.8))
```

---

*There is one exception: the `grid.arrows()` function makes use of the current location
when an arrow is added to a `line.to` graphical object produced by `lineToGrob()`.

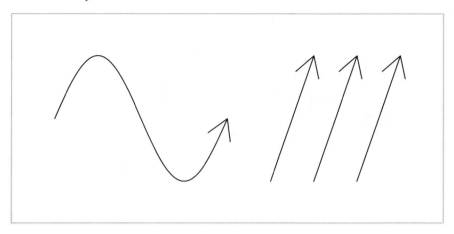

Figure 5.3
Drawing arrows using the `grid.arrows()` function. Arrows can be added to: a single line through multiple points, as generated by `grid.lines()` (e.g., the sine curve in the left half of the figure); multiple straight line segments, as generated by `grid.segments()` (e.g., the three straight lines in the right half of the figure); the result of a line-to operation, as generated by `grid.line.to()` (example not shown here).

In simple usage, the `grid.polygon()` function draws a single polygon through the specified x and y locations (automatically joining the last location to the first to close the polygon). It is possible to produce multiple polygons from a single call (which is much faster than making multiple calls) if the `id` argument is specified. In this case, a polygon is drawn for each set of x and y locations corresponding to a different value of `id`. The following code demonstrates both usages (see Figure 5.4). The two `grid.polygon()` calls use the same x and y locations, but the second call splits the locations into three separate polygons using the `id` argument.

```
> angle <- seq(0, 2*pi, length=10)[-10]
> grid.polygon(x=0.25 + 0.15*cos(angle), y=0.5 + 0.3*sin(angle),
               gp=gpar(fill="grey"))
> grid.polygon(x=0.75 + 0.15*cos(angle), y=0.5 + 0.3*sin(angle),
               id=rep(1:3, each=3),
               gp=gpar(fill="grey"))
```

The `grid.xaxis()` and `grid.yaxis()` functions are not really graphical primitives as they produce relatively complex output consisting of both lines and text. They are included here because they complete the set of grid functions that produce graphical output. The main argument to these functions is the

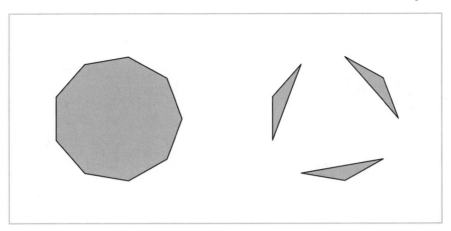

Figure 5.4
Drawing polygons using the `grid.polygon()` function. By default, a single polygon is produced from multiple (`x, y`) locations (the nonagon on the left), but it is possible to associate subsets of the locations with separate polygons using the `id` argument (the three triangles on the right).

`at` argument. This is used to specify where tick-marks should be placed. If the argument is not specified, sensible tick-marks are drawn based on the current scales in effect (see Section 5.5 for information about viewport scales). The values specified for the `at` argument are always relative to the current scales (see the concept of the `"native"` coordinate system in Section 5.3). These functions are much less flexible and general than the traditional `axis()` function. For example, they do not provide automatic support for generating labels from time- or date-based `at` locations.

Drawing curves

There is no native curve-drawing function in grid, but an approximation to a smooth curve consisting of many straight line segments is often sufficient. The example on the left of Figure 5.3 demonstrates how a series of line segments can appear very much like a smooth curve, if enough line segments are used.

5.2.1   Standard arguments

All primitive graphics functions accept a `gp` argument that allows control over aspects such as the color and line type of the relevant output. For example, the following code specifies that the boundary of the rectangle should be dashed and colored red.

```
> grid.rect(gp=gpar(col="red", lty="dashed"))
```

Section 5.4 provides more information about setting graphical parameters.

All primitive graphics functions also accept a vp argument that can be used to specify a viewport in which to draw the relevant output. The following code shows a simple example of the syntax (the result is a rectangle drawn in the left half of the page); Section 5.5 describes viewports and the use of vp arguments in full detail.

```
> grid.rect(vp=viewport(x=0, width=0.5, just="left"))
```

Finally, all primitive graphics functions also accept a name argument. This can be used to identify the graphical object produced by the function. It is useful for interactively editing graphical output and when working with graphical objects (see Chapter 6). The following code demonstrates how to associate a name with a rectangle.

```
> grid.rect(name="myrect")
```

5.3 Coordinate systems

When drawing in grid, there are always a large number of coordinate systems available for specifying the locations and sizes of graphical output. For example, it is possible to specify an x location as a proportion of the width of the drawing region, or as a number of inches (or centimeters, or millimeters) from the left-hand edge of the drawing region, or relative to the current x-scale. The full set of coordinate systems available is shown in Table 5.2. The meaning of some of these will only become clear with an understanding of viewports (Section 5.5) and graphical objects (Chapter 6).*

With so many coordinate systems available, it is necessary to specify which coordinate system a location or size refers to. The unit() function is used

---

*Absolute units, such as inches, may not be rendered with full accuracy on screen devices (see the footnote on page 100).

Table 5.2

The full set of coordinate systems available in grid.

Coordinate System Name	Description
`"native"`	Locations and sizes are relative to the x- and y-scales for the current viewport.
`"npc"`	Normalized Parent Coordinates. Treats the bottom-left corner of the current viewport as the location $(0, 0)$ and the top-right corner as $(1, 1)$.
`"snpc"`	Square Normalized Parent Coordinates. Locations and sizes are expressed as a proportion of the *smaller* of the width and height of the current viewport.
`"inches"`	Locations and sizes are in terms of physical inches. For locations, $(0, 0)$ is at the bottom-left of the viewport.
`"cm"`	Same as `"inches"`, except in centimeters.
`"mm"`	Millimeters.
`"points"`	Points. There are 72.27 points per inch.
`"bigpts"`	Big points. There are 72 big points per inch.
`"picas"`	Picas. There are 12 points per pica.
`"dida"`	Dida. 1157 dida equals 1238 points.
`"cicero"`	Cicero. There are 12 dida per cicero.
`"scaledpts"`	Scaled points. There are 65536 scaled points per point.
`"char"`	Locations and sizes are specified in terms of multiples of the current nominal font size (dependent on the current `fontsize` and `cex`).
`"lines"`	Locations and sizes are specified in terms of multiples of the height of a line of text (dependent on the current `fontsize`, `cex`, and `lineheight`).
`"strwidth"` `"strheight"`	Locations and sizes are expressed as multiples of the width (or height) of a given string (dependent on the string and the current `fontsize`, `cex`, `fontfamily`, and `fontface`).
`"grobwidth"` `"grobheight"`	Locations and sizes are expressed as multiples of the width (or height) of a given graphical object (dependent on the type, location, and graphical settings of the graphical object).

to associate a numeric value with a coordinate system. This function creates an object of class `"unit"` (hereafter referred to simply as a *unit*), which acts very much like a normal `numeric` object — it is possible to perform basic operations such as sub-setting units, and adding and subtracting units.

Each value in a unit can be associated with a different coordinate system and each location and dimension of a graphical object is a separate unit, so for example, a rectangle can have its x-location, y-location, width, and height all specified relative to different coordinate systems.

The following pieces of code demonstrate some of the flexibility of grid units. The first code examples show some different uses of the `unit()` function: a single value is associated with a coordinate system, then several values are associated with a coordinate system (notice the recycling of the coordinate system value), then several values are associated with different coordinate systems.

```
> unit(1, "mm")
```

```
[1] 1mm
```

```
> unit(1:4, "mm")
```

```
[1] 1mm 2mm 3mm 4mm
```

```
> unit(1:4, c("npc", "mm", "native", "lines"))
```

```
[1] 1npc    2mm    3native 4lines
```

The next code examples show how units can be manipulated in many of the ways that normal numeric vectors can: firstly by sub-setting, then simple addition (again notice the recycling), then finally the use of a summary function (`max()` in this case).

```
> unit(1:4, "mm")[2:3]
```

```
[1] 2mm 3mm
```

```
> unit(1, "npc") - unit(1:4, "mm")
```

```
[1] 1npc-1mm 1npc-2mm 1npc-3mm 1npc-4mm
```

```
> max(unit(1:4, c("npc", "mm", "native", "lines")))
```

[1] max(1npc, 2mm, 3native, 4lines)

Some operations on units are not as straightforward as with numeric vectors, but require the use of functions written specifically for units. For example, the length of units must be obtained using the `unit.length()` function rather than `length()`, units must be concatenated (in the sense of the `c()` function) using `unit.c()`, and there are special functions for repeating units and for calculating parallel maxima and minima (`unit.rep()`, `unit.pmin()`, and `unit.pmax()`).

The following code provides an example of using units to locate and size a rectangle. The rectangle is at a location 40% of the way across the drawing region and 1 inch from the bottom of the drawing region. It is as wide as the text `"very snug"`, and it is one line of text high (see Figure 5.5).

```
> grid.rect(x=unit(0.4, "npc"), y=unit(1, "inches"),
            width=stringWidth("very snug"),
            height=unit(1, "lines"),
            just=c("left", "bottom"))
```

5.3.1   Conversion functions

As demonstrated in the previous section, a unit is not simply a numeric value. Units only reduce to a simple numeric value (a physical location on a graphics device) when drawing occurs. A consequence of this is that a unit can mean very different things, depending on when it gets drawn (this should become more apparent with an understanding of graphical parameters in Section 5.4 and viewports in Section 5.5).

In some cases, it can be useful to convert a unit to a simple numeric value. For example, it is sometimes necessary to know the current scale limits for numerical calculations. There are several functions that can assist with this problem: `convertUnit()`, `convertX()`, `convertY()`, `convertWidth()`, and `convertHeight()`. The following code demonstrates how to calculate the current scale limits for the x-dimension. First of all, a scale is defined on the x-axis with the range `c(-10, 50)` (see Section 5.5 for more about viewports).

```
> pushViewport(viewport(xscale=c(-10, 50)))
```

The next expression performs a query to obtain the current x-axis scale. The expression `unit(0:1, "npc")` represents the left and right boundaries of the

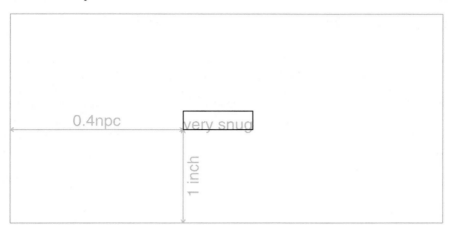

Figure 5.5
A demonstration of grid units. A diagram demonstrating how graphical output
can be located and sized using grid units to associate numeric values with different
coordinate systems. The grey border represents the current viewport. A black
rectangle has been drawn with its bottom-left corner 40% of the way across the
current viewport and 1 inch above the bottom of the current viewport. The rectangle
is 1 line of text high and as wide as the text "very snug" (as it would be drawn in
the current font).

current drawing region and `convertX()` is used to convert these locations into
values in the `"native"` coordinate system, which is relative to the current
scales.

```
> convertX(unit(0:1, "npc"), "native", valueOnly=TRUE)
```

```
[1] -10  50
```

WARNING: These conversion functions must be used with care. The output from these functions is only valid for the current device size. If, for
example, a window on screen is resized, or output is copied from one device to
another device with a different physical size, these calculations may no longer
be correct. In other words, only rely on these functions when it is known
that the size of the graphics device will not change. See Appendix B for more
information on this topic and for a way to be able to use these functions on
devices that may be resized. The discussion on the use of these functions
in `drawDetails()` methods and the function `grid.record()` is also relevant
(see "Calculations during drawing" in Section 7.3.10).

5.3.2   Complex units

There are two peculiarities of the "strwidth", "strheight", "grobwidth",
and "grobheight" coordinate systems that require further explanation. In
all of these cases, a value is interpreted as a multiple of the size of some
other object. In the former two cases, the other object is just a text string
(e.g., "a label"), but in the latter two cases, the other object can be any
graphical object (see Chapter 6). It is necessary to specify the other object
when generating a unit for these coordinate systems and this is achieved via
the data argument. The following code shows some simple examples.

```
> unit(1, "strwidth", "some text")
```

```
[1] 1strwidth
```

```
> unit(1, "grobwidth", textGrob("some text"))
```

```
[1] 1grobwidth
```

A more convenient interface for generating units, when all values are relative to a single coordinate system, is also available via the stringWidth(),
stringHeight(), grobWidth(), and grobHeight() functions. The following
code is equivalent to the previous example.

```
> stringWidth("some text")
```

```
[1] 1strwidth
```

```
> grobWidth(textGrob("some text"))
```

```
[1] 1grobwidth
```

In this particular example, the "strwidth" and "grobwidth" units will be
identical as they are based on identical pieces of text. The difference is that
a graphical object can contain not only the text to draw, but other information that may affect the size of the text, such as the font family and size.
In the following code, the two units are no longer identical because the text
grob represents text drawn at font size of 18, whereas the simple string represents text at the default size of 10. The convertWidth() function is used
to demonstrate the difference.

```
> convertWidth(stringWidth("some text"), "inches")
```

[1] 0.7175inches

```
> convertWidth(grobWidth(textGrob("some text",
                                  gp=gpar(fontsize=18))),
              "inches")
```

[1] 1.07625inches

For units that contain multiple values, there must be an object specified for every "strwidth", "strheight", "grobwidth", and "grobheight" value. Where there is a mixture of coordinate systems within a unit, a value of NULL can be supplied for the coordinate systems that do not require data. The following code demonstrates this.

```
> unit(rep(1, 3), "strwidth", list("one", "two", "three"))
```

[1] 1strwidth 1strwidth 1strwidth

```
> unit(rep(1, 3),
       c("npc", "strwidth", "grobwidth"),
       list(NULL, "two", textGrob("three")))
```

[1] 1npc   1strwidth   1grobwidth

Again, there is a simpler interface for straightforward situations.

```
> stringWidth(c("one", "two", "three"))
```

[1] 1strwidth 1strwidth 1strwidth

For "grobwidth" and "grobheight" units, it is also possible to specify the name of a graphical object rather than the graphical object itself. This can be useful for establishing a reference to a graphical object, so that when the named graphical object is modified, the unit is updated for the change. The following code demonstrates this idea. First of all, a text grob is created with the name "tgrob".

```
> grid.text("some text", name="tgrob")
```

Next, a unit is created that is based on the width of the grob called `"tgrob"`.

```
> theUnit <- grobWidth("tgrob")
```

The `convertWidth()` function can be used to show the current value of the unit.

```
> convertWidth(theUnit, "inches")
```

```
[1] 0.7175inches
```

The following code modifies the grob named `"tgrob"` and `convertWidth()` is used to show that the value of the unit reflects the new width of the `text` grob.

```
> grid.edit("tgrob", gp=gpar(fontsize=18))
> convertWidth(theUnit, "inches")
```

```
[1] 1.07625inches
```

5.4   Controlling the appearance of output

All graphical primitives functions (and the `viewport()` function — see Section 5.5) — have a `gp` argument that can be used to provide a set of graphical parameters to control the appearance of the graphical output. There is a fixed set of graphical parameters (see Table 5.3), all of which can be specified for all types of graphical output.

The value supplied for the `gp` argument must be an object of class `"gpar"`, and a `gpar` object can be produced using the `gpar()` function. For example, the following code produces a `gpar` object containing graphical parameter settings controlling color and line type.

```
> gpar(col="red", lty="dashed")
```

```
$col
[1] "red"

$lty
[1] "dashed"
```

Table 5.3

The full set of graphical parameters available in grid. The `lex` parameter has only been available since R version 2.1.0.

Parameter	Description
`col`	Color of lines, text, rectangle borders, ...
`fill`	Color for filling rectangles, circles, polygons, ...
`gamma`	Gamma correction for colors
`alpha`	Alpha blending coefficient for transparency
`lwd`	Line width
`lex`	Line width expansion multiplier applied to `lwd` to obtain final line width
`lty`	Line type
`lineend`	Line end style (round, butt, square)
`linejoin`	Line join style (round, mitre, bevel)
`linemitre`	Line mitre limit
`cex`	Character expansion multiplier applied to `fontsize` to obtain final font size
`fontsize`	Size of text (in points)
`fontface`	Font face (bold, italic, ...)
`fontfamily`	Font family
`lineheight`	Multiplier applied to final font size to obtain the height of a line

The function `get.gpar()` can be used to obtain current graphical parameter settings. The following code shows how to query the current line type and fill color. When called with no arguments, the function returns a complete list of current settings.

```
> get.gpar(c("lty", "fill"))


$lty
[1] "solid"

$fill
[1] "transparent"
```

A `gpar` object represents an *explicit graphical context* — settings for a small number of specific graphical parameters. The example above produces a graphical context that ensures that the color setting is `"red"` and the line-type setting is `"dashed"`. There is always an *implicit graphical context* consisting of default settings for all graphical parameters. The implicit graphical context is initialized automatically by grid for every graphics device and can be modified by viewports (see Section 5.5.5) or by gTrees (see Section 6.2.1).*

A graphical primitive will be drawn with graphical parameter settings taken from the implicit graphical context, except where there are explicit graphical parameter settings from the graphical primitive's `gp` argument. For graphical primitives, the explicit graphical context is only in effect for the duration of the drawing of the graphical primitive. The following code example demonstrates these rules.

The default initial implicit graphical context includes settings such as `lty="solid"` and `fill="transparent"`. The first (left-most) rectangle has an explicit setting `fill="black"` so it only uses the implicit setting `lty="solid"`. The second (right-most) rectangle uses all of the implicit graphical parameter settings. In particular, it is not at all affected by the explicit settings of the first rectangle (see Figure 5.6).

```
> grid.rect(x=0.66, height=0.7, width=0.2,
            gp=gpar(fill="black"))
> grid.rect(x=0.33, height=0.7, width=0.2)
```

---

*The ideas of implicit and explicit graphical contexts are similar to the specification of settings in Cascading Style Sheets[34] and the graphics state in PostScript[3].

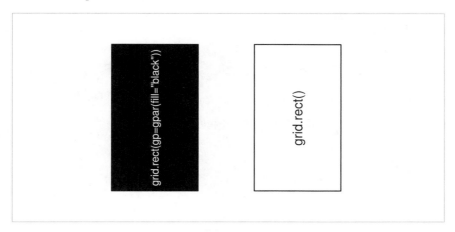

Figure 5.6
Graphical parameters for graphical primitives. The grey rectangle represents the current viewport. The right-hand rectangle has been drawn with no specific graphical parameters so it inherits the defaults for the current viewport (which in this case are a black border and no fill color). The left-hand rectangle has been drawn with a specific fill color of black (it is still drawn with the inherited black border). The graphical parameter settings for one rectangle have no effect on the other rectangle.

5.4.1 Specifying graphical parameter settings

The values that can be specified for colors, line types, line widths, line ends, line joins, and fonts are mostly the same as for the traditional graphics system. Sections 3.2.1, 3.2.2, and 3.2.3 contain descriptions of these specifications (for example, see the sub-section "Specifying colors"). In many cases, the graphical parameter in grid also has the same name as the traditional graphics state setting (e.g., `col`), though several of the grid parameters are slightly more verbose (e.g. `lineend` and `fontfamily`). Some other differences in the specification of graphical parameter values in the grid graphics system are described below.

In grid, the `fontface` value can be a string instead of an integer. Table 5.4 shows the possible string values.

In grid, the `cex` value is cumulative. This means that it is multiplied by the previous `cex` value to obtain a current `cex` value. The following code shows a simple example. A viewport is pushed with `cex=0.5`. This means that text will be half size. Next, some text is drawn, also with `cex=0.5`. This text is drawn quarter size because `cex` was already `0.5` from the viewport (`0.5*0.5 = 0.25`).

Table 5.4
Possible font face specifications in grid.

Integer	String	Description
1	"plain"	Roman or upright face
2	"bold"	Bold face
3	"italic" or "oblique"	Slanted face
4	"bold.italic"	Bold and slanted face
For the HersheySerif font family		
5	"cyrillic"	Cyrillic font
6	"cyrillic.oblique"	Slanted Cyrillic font
7	"EUC"	Japanese characters

```
> pushViewport(viewport(gp=gpar(cex=0.5)))
> grid.text("How small do you think?", gp=gpar(cex=0.5))
```

The **alpha** graphical parameter setting is unique to grid. It is a value between 1 (fully opaque) and 0 (fully transparent). The **alpha** value is combined with the alpha channel of colors by multiplying the two and this setting is cumulative like the **cex** setting. The following code shows a simple example. A viewport is pushed with **alpha=0.5**, then a rectangle is drawn using a semitransparent red fill color (alpha channel set to 0.5). The final alpha channel for the fill color is 0.25 (0.5*0.5 = 0.25).

```
> pushViewport(viewport(gp=gpar(alpha=0.5)))
> grid.rect(width=0.5, height=0.5,
            gp=gpar(fill=rgb(1, 0, 0, 0.5)))
```

Grid does not support fill patterns (see page 58).

5.4.2　Vectorized graphical parameter settings

All graphical parameter settings may be vector values. Many graphical primitive functions produce multiple primitives as output and graphical parameter settings will be recycled over those primitives. The following code produces 100 circles, cycling through 50 different shades of grey for the circles (see Figure 5.7).

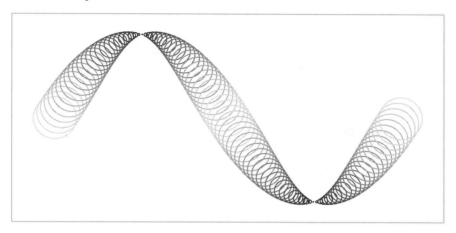

Figure 5.7
Recycling graphical parameters. The 100 circles are drawn by a single function call with 50 different greys specified for the border color (from a very light grey to a very dark grey and back to a very light grey). The 50 colors are recycled over the 100 circles so circle i gets the same color as circle $i + 50$.

```
> levels <- round(seq(90, 10, length=25))
> greys <- paste("grey", c(levels, rev(levels)), sep="")
> grid.circle(x=seq(0.1, 0.9, length=100),
              y=0.5 + 0.4*sin(seq(0, 2*pi, length=100)),
              r=abs(0.1*cos(seq(0, 2*pi, length=100))),
              gp=gpar(col=greys))
```

The grid.polygon() function is a slightly complex case. There are two ways in which this function will produce multiple polygons: when the id argument is specified *and* when there are NA values in the x or y locations (see Section 5.6). For grid.polygon(), a different graphical parameter will only be applied to each polygon identified by a different id. When a single polygon (as identified by a single id value) is split into multiple sub-polygons by NA values, all sub-polygons receive the same graphical parameter settings. The following code demonstrates these rules (see Figure 5.8). The first call to grid.polygon() draws two polygons as specified by the id argument. The fill graphical parameter setting contains two colors so the first polygon gets the first color (grey) and the second polygon gets the second color (white). In the second call, all that has changed is that an NA value has been introduced. This means that the first polygon as specified by the id argument is split into two separate polygons, but both of these polygons use the same fill setting because they both correspond to an id of 1. Both of these polygons get the first color (grey).

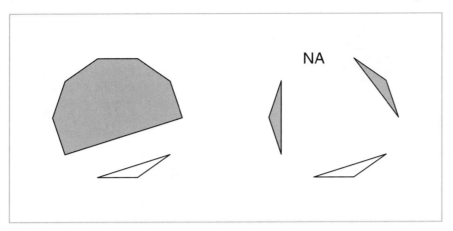

Figure 5.8
Recycling graphical parameters for polygons. On the left, a single function call
produces two polygons with different fill colors by specifying an id argument and
two fill colors. On the right, there are three polygons because an NA value has been
introduced in the (x, y) locations for the polygon, but there are still only two colors
specified. The colors are allocated to polygons using the id argument and ignoring
any NA values.

```
> angle <- seq(0, 2*pi, length=11)[-11]
> grid.polygon(x=0.25 + 0.15*cos(angle), y=0.5 + 0.3*sin(angle),
               id=rep(1:2, c(7, 3)),
               gp=gpar(fill=c("grey", "white")))
> angle[4] <- NA
> grid.polygon(x=0.75 + 0.15*cos(angle), y=0.5 + 0.3*sin(angle),
               id=rep(1:2, c(7, 3)),
               gp=gpar(fill=c("grey", "white")))
```

All graphical primitives have a gp component, so it is possible to specify any
graphical parameter setting for any graphical primitive. This may seem inef-
ficient, and indeed in some cases the values are completely ignored (e.g., text
drawing ignores the lty setting), but in many cases the values are potentially
useful. For example, even when there is no text being drawn, the settings for
fontsize, cex, and lineheight are always used to calculate the meaning of
"lines" and "char" coordinates.

5.5   Viewports

A *viewport* is a rectangular region that provides a context for drawing.

A viewport provides a *drawing context* consisting of both a *geometric context* and a *graphical context*. A geometric context consists of a set of coordinate systems for locating and sizing output and all of the coordinate systems described in Section 5.3 are available within every viewport.* A graphical context consists of explicit graphical parameter settings for controlling the appearance of output. This is specified as a `gpar` object via the `gp` argument.

By default, grid creates a viewport that corresponds to the entire graphics device and, until another viewport is created, drawing occurs within the full extent of the device and using the default graphical parameter settings.

A new viewport is created using the `viewport()` function. A viewport has a location (given by `x` and `y`), a size (given by `width` and `height`), and it is justified relative to its location (according to the value of the `just` argument). The location and size of a viewport are specified in units, so a viewport can be positioned and sized within another viewport in a very flexible manner. The following code creates a viewport that is left-justified at an `x` location 0.4 of the way across the drawing region, and bottom-justified 1 centimeter from the bottom of the drawing region. It is as wide as the text `"very very snug indeed"`, and it is six lines of text high. Figure 5.9 shows a diagram representing this viewport.

```
> viewport(x=unit(0.4, "npc"), y=unit(1, "cm"),
           width=stringWidth("very very snug indeed"),
           height=unit(6, "lines"),
           just=c("left", "bottom"))
```

viewport [GRID.VP.33]

An important thing to notice in the above example is that the result of the `viewport()` function is an object of class `viewport`. No region has actually been created on a graphics device. In order to create regions on a graphics device, a `viewport` object must be *pushed* onto the device, as described in the next section.

---

*The idea of being able to define a geometric context is similar to the concept of the current transformation matrix (CTM) in PostScript[3] and the modeling transformation in OpenGL[55].

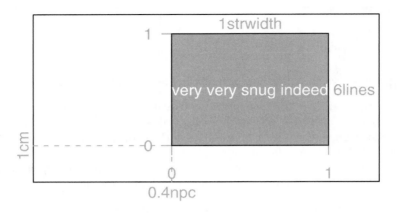

Figure 5.9
A diagram of a simple viewport. A viewport is a rectangular region specified by
an (x, y) location, a (width, height) size, and a justification (and possibly a
rotation). This diagram shows a viewport that is left-bottom justified 1 centimeter
off the bottom of the page and 0.4 of the way across the page. It is 6 lines of text
high and as wide as the text "very very snug indeed".

5.5.1 Pushing, popping, and navigating between viewports

The pushViewport() function takes a viewport object and uses it to create
a region on the graphics device. This region becomes the drawing context for
all subsequent graphical output, until the region is removed or another region
is defined.

The following code demonstrates this idea (see Figure 5.10). To start with,
the entire device, and the default graphical parameter settings, provide the
drawing context. Within this context, the grid.text() call draws some text
at the top-left corner of the device. A viewport is then pushed, which creates
a region 80% as wide as the device, half the height of the device, and rotated
at an angle of 10 degrees*. The viewport is given a name, "vp1", which will
help us to navigate back to this viewport from another viewport later.

Within the new drawing context defined by the viewport that has been pushed,
exactly the same grid.text() call produces some text at the top-left corner
of the viewport. A rectangle is also drawn to make the extent of the new
viewport clear.

---

*It is not often very useful to rotate a viewport, but it helps in this case to dramatise
the difference between the drawing regions.

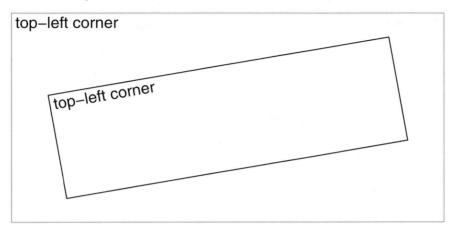

Figure 5.10
Pushing a viewport. Drawing occurs relative to the entire device until a viewport is
pushed. For example, some text has been drawn in the top-left corner of the device.
Once a viewport has been pushed, output is drawn relative to that viewport. The
black rectangle represents a viewport that has been pushed and text has been drawn
in the top-left corner of that viewport.

```
> grid.text("top-left corner", x=unit(1, "mm"),
            y=unit(1, "npc") - unit(1, "mm"),
            just=c("left", "top"))
> pushViewport(viewport(width=0.8, height=0.5, angle=10,
                name="vp1"))
> grid.rect()
> grid.text("top-left corner", x=unit(1, "mm"),
            y=unit(1, "npc") - unit(1, "mm"),
            just=c("left", "top"))
```

The pushing of viewports is entirely general. A viewport is pushed relative
to the current drawing context. The following code slightly extends the previous example by pushing a further viewport, exactly like the first, and again
drawing text at the top-left corner (see Figure 5.11). The location, size, and
rotation of this second viewport are all relative to the context provided by the
first viewport. Viewports can be nested like this to any depth.

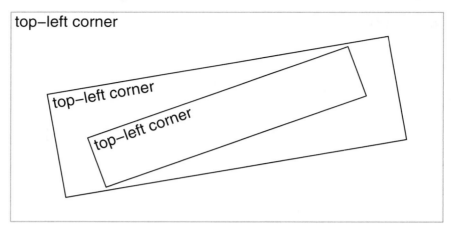

Figure 5.11
Pushing several viewports. Viewports are pushed relative to the current viewport.
Here, a second viewport has been pushed relative to the viewport that was pushed
in Figure 5.10. Again, text has been drawn in the top-left corner.

```
> pushViewport(viewport(width=0.8, height=0.5, angle=10,
                name="vp2"))
> grid.rect()
> grid.text("top-left corner", x=unit(1, "mm"),
            y=unit(1, "npc") - unit(1, "mm"),
            just=c("left", "top"))
```

In grid, drawing is always within the context of the current viewport. One
way to change the current viewport is to push a viewport (as in the previous
examples), but there are other ways too. For a start, it is possible to *pop* a
viewport using the `popViewport()` function. This removes the current viewport and the drawing context reverts to whatever it was before the current
viewport was pushed*. The following code demonstrates popping viewports
(see Figure 5.12). The call to `popViewport()` removes the last viewport created on the device. Text is drawn at the bottom-right of the resulting drawing
region (which has reverted back to being the first viewport that was pushed).

```
> popViewport()
> grid.text("bottom-right corner",
            x=unit(1, "npc") - unit(1, "mm"),
            y=unit(1, "mm"), just=c("right", "bottom"))
```

---

*It is illegal to pop the top-most viewport that represents the entire device region and
the default graphical parameter settings. Trying to do so will result in an error.

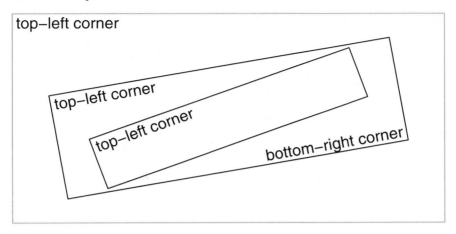

Figure 5.12
Popping a viewport. When a viewport is popped, the drawing context reverts to
the parent viewport. In this figure, the second viewport (pushed in Figure 5.11) has
been popped to go back to the first viewport (pushed in Figure 5.10). This time
text has been drawn in the bottom-right corner.

The popViewport() function has an integer argument n that specifies how
many viewports to pop. The default is 1, but several viewports can be popped
at once by specifying a larger value. The special value of 0 means that all
viewports should be popped. In other words, the drawing context should
revert to the entire device and the default graphical parameter settings.

Another way to change the current viewport is by using the upViewport()
and downViewport() functions. The upViewport() function is similar to
popViewport() in that the drawing context reverts to whatever it was prior to
the current viewport being pushed. The difference is that upViewport() does
not remove the current viewport from the device. This difference is significant
because it means that that a viewport can be revisited without having to push
it again. Revisiting a viewport is faster than pushing a viewport and it allows
the creation of viewport regions to be separated from the production of output
(see "viewport paths" in Section 5.5.3 and Chapter 7).

A viewport can be revisited using the downViewport() function. This function
has an argument name that can be used to specify the name of an existing
viewport. The result of downViewport() is to make the named viewport
the current drawing context. The following code demonstrates the use of
upViewport() and downViewport() (see Figure 5.13).

A call to upViewport() is made, which reverts the drawing context to the
entire device (recall that prior to this navigation the current viewport was
the first viewport that was pushed) and text is drawn in the bottom-right

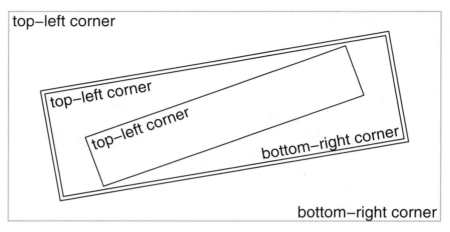

Figure 5.13
Navigating between viewports. Rather than popping a viewport, it is possible to
navigate up from a viewport (and leave the viewport on the device). Here navigation
has occurred from the first viewport to revert the drawing context to the entire
device and text has been drawn in the bottom-right corner. Next, there has been
a navigation down to the first viewport again and a second border has been drawn
around the outside of the viewport.

corner. The downViewport() function is then used to navigate back down to
the viewport that was first pushed and a second border is drawn around this
viewport. The viewport to navigate down to is specified by its name, "vp1".

```
> upViewport()
> grid.text("bottom-right corner",
            x=unit(1, "npc") - unit(1, "mm"),
            y=unit(1, "mm"), just=c("right", "bottom"))
> downViewport("vp1")
> grid.rect(width=unit(1, "npc") + unit(2, "mm"),
            height=unit(1, "npc") + unit(2, "mm"))
```

There is also a seekViewport() function that can be used to travel across
the viewport tree. This can be convenient for interactive use, but the result is
less predictable, so it is less suitable for use in writing grid functions for others to use. The call seekViewport("avp") is equivalent to upViewport(0);
downViewport("avp").

Drawing between viewports

Sometimes it is useful to be able to locate graphical output relative to more than one viewport. The only way to do this in grid is via the `grid.move.to()` and `grid.line.to()` functions. It is possible to call `grid.move.to()` within one viewport, change viewports, and call `grid.line.to()`. An example is provided in Section 5.8.2.

5.5.2   Clipping to viewports

Drawing can be restricted to only the interior of the current viewport (*clipped* to the viewport) by specifying the `clip` argument to the `viewport()` function. This argument has three values: `"on"` indicates that output should be clipped to the current viewport; `"off"` indicates that output should not be clipped at all; `"inherit"` means that the clipping region of the previous viewport should be used (this may not have been set by the previous viewport if that viewport's `clip` argument was also `"inherit"`). The following code provides a simple example (see Figure 5.14). A viewport is pushed with clipping on and a circle with a very thick black border is drawn relative to the viewport. A rectangle is also drawn to show the extent of the viewport. The circle partially extends beyond the limits of the viewport, so only those parts of the circle that lie within the viewport are drawn.

```
> pushViewport(viewport(w=.5, h=.5, clip="on"))
> grid.rect()
> grid.circle(r=.7, gp=gpar(lwd=20))
```

Next, another viewport is pushed and this viewport just inherits the clipping region from the first viewport. Another circle is drawn, this time with a grey and slightly thinner border and again the circle is clipped to the viewport.

```
> pushViewport(viewport(clip="inherit"))
> grid.circle(r=.7, gp=gpar(lwd=10, col="grey"))
```

Finally, a third viewport is pushed with clipping turned off. Now, when a third circle is drawn (with a thin, black border) all of the circle is drawn, even though parts of the circle extend beyond the the viewport.

```
> pushViewport(viewport(clip="off"))
> grid.circle(r=.7)
> popViewport(3)
```

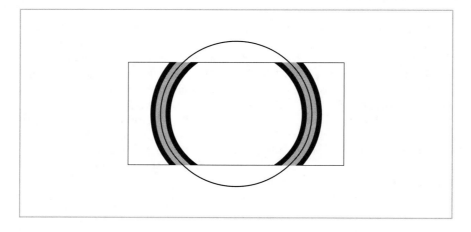

Figure 5.14

Clipping output in viewports. When a viewport is pushed, output can be clipped to
that viewport, or the clipping region can be left in its current state, or clipping can
be turned off entirely. In this figure, a viewport is pushed (the black rectangle) with
clipping on. A circle is drawn with a very thick black border and it gets clipped.
Next, another viewport is pushed (in the same location) with clipping left as it was.
A second circle is drawn with a slightly thinner grey border and it is also clipped.
Finally, a third viewport is pushed, which turns clipping off. A circle is drawn with
a thin black border and this circle is not clipped.

5.5.3 Viewport lists, stacks, and trees

It can be convenient to work with several viewports at once and there are several facilities for doing this in grid. The `pushViewport()` function will accept multiple arguments and will push the specified viewports one after another. For example, the fourth expression below is a shorter equivalent version of the first three expressions.

```
> pushViewport(vp1)
> pushViewport(vp2)
> pushViewport(vp3)
```

```
> pushViewport(vp1, vp2, vp3)
```

The `pushViewport()` function will also accept objects that contain several viewports: viewport lists, viewport stacks, and viewport trees. The function `vpList()` creates a list of viewports and these are pushed "in parallel." The first viewport in the list is pushed, then grid navigates back up before the next viewport in the list is pushed. The `vpStack()` function creates a stack of viewports and these are pushed "in series." Pushing a stack of viewports is exactly the same as specifying the viewports as multiple arguments to `pushViewport()`. The `vpTree()` function creates a tree of viewports that consists of a parent viewport and any number of child viewports. The parent viewport is pushed first, then the the child viewports are pushed in parallel within the parent.

The current set of viewports that have been pushed on the current device constitute a viewport tree and the `current.vpTree()` function prints out a representation of the current viewport tree. The following code demonstrates the output from `current.vpTree()` and the difference between lists, stacks, and trees of viewports. First of all, some (trivial) viewports are created to work with.

```
> vp1 <- viewport(name="A")
> vp2 <- viewport(name="B")
> vp3 <- viewport(name="C")
```

The next piece of code shows these three viewports pushed as a list. The output of `current.vpTree()` shows the root viewport (which represents the entire device) and then all three viewports as children of the root viewport.

```
> pushViewport(vpList(vp1, vp2, vp3))
> current.vpTree()
```

```
viewport[ROOT]->(viewport[A], viewport[B], viewport[C])
```

This next code pushes the three viewports as a stack. The viewport **vp1** is now the only child of the root viewport with **vp2** a child of **vp1**, and **vp3** a child of **vp2**.

```
> grid.newpage()
> pushViewport(vpStack(vp1, vp2, vp3))
> current.vpTree()
```

```
viewport[ROOT]->(viewport[A]->(viewport[B]->(viewport[C])))
```

Finally, the three viewports are pushed as a tree, with **vp1** as the parent and **vp2** and **vp3** as its children.

```
> grid.newpage()
> pushViewport(vpTree(vp1, vpList(vp2, vp3)))
> current.vpTree()
```

```
viewport[ROOT]->(viewport[A]->(viewport[B], viewport[C]))
```

As with single viewports, viewport lists, stacks, and trees can be provided as the **vp** argument for graphical functions (see Section 5.5.4).

Viewport paths

The `downViewport()` function, by default, searches down the current viewport tree as far as is necessary to find a given viewport name. This is convenient for interactive use, but can be ambiguous if there is more than one viewport with the same name in the viewport tree.

Grid provides the concept of a *viewport path* to resolve such ambiguity. A viewport path is an ordered list of viewport names, which specify a series of parent-child relations. A viewport path is created using the `vpPath()` function. For example, the following code produces a viewport path that specifies a viewport called `"C"` with a parent called `"B"`, which in turn has a parent called `"A"`.

```
> vpPath("A", "B", "C")
```

```
A::B::C
```

For convenience in interactive use, a viewport path may be specified directly as a string. For example, the previous viewport path could be specified simply as `"A::B::C"`. The `vpPath()` function should be used when writing graphics functions for others to use.

The `name` argument to the `downViewport()` function will accept a viewport path, in which case it searches for a viewport that matches the entire path. The `strict` argument to `downViewport()` ensures that a viewport will only be found if the full viewport path is found, *starting from the current location in the viewport tree.*

5.5.4 Viewports as arguments to graphical primitives

As mentioned in Section 5.2.1, a viewport may be specified as an argument to functions that produce graphical output (via an argument called **vp**). When a viewport is specified in this way, the viewport gets pushed before the graphical output is produced and popped afterwards. To make this completely clear, the following two code segments are identical. First of all, a simple viewport is defined.

```
> vp1 <- viewport(width=0.5, height=0.5, name="vp1")
```

The next code explicitly pushes the viewport, draws some text, then pops the viewport.

```
> pushViewport(vp1)
> grid.text("Text drawn in a viewport")
> popViewport()
```

This next piece of code does the same thing in a single call.

```
> grid.text("Text drawn in a viewport", vp=vp1)
```

It is also possible to specify the name of a viewport (or a viewport path) for a vp argument. In this case, the name (or path) is used to navigate down to the viewport (via a call to `downViewport()`) and then back up again afterwards (via a call to `upViewport()`). This promotes the practice of pushing viewports once, then specifying where to draw different output by simply naming the appropriate viewport. The following code does the same thing as the previous example, but leaves the viewport intact (so that it can be used for further drawing).

```
> pushViewport(vp1)
> upViewport()
> grid.text("Text drawn in a viewport", vp="vp1")
```

This feature is also very useful when annotating a plot produced by a high-level graphics function. As long as the graphics function names the viewports that it creates and does not pop them, it is possible to revisit the viewports to add further output. Examples of this are given in Section 5.8 and this approach to writing high-level grid functions is discussed further in Chapter 7.

5.5.5 Graphical parameter settings in viewports

A viewport can have graphical parameter settings associated with it via the gp argument to viewport(). When a viewport has graphical parameter settings, those settings affect all graphical objects drawn within the viewport, and all other viewports pushed within the viewport, unless the graphical objects or the other viewports specify their own graphical parameter setting. In other words, the graphical parameter settings for a viewport modify the implicit graphical context (see page 168).

The following code demonstrates this rule. A viewport is pushed that has a fill="grey" setting. A rectangle with no graphical parameter settings is drawn within that viewport and this rectangle "inherits" the fill="grey" setting. Another rectangle is drawn with its own fill setting so it does not inherit the viewport setting (see Figure 5.15).

```
> pushViewport(viewport(gp=gpar(fill="grey")))
> grid.rect(x=0.33, height=0.7, width=0.2)
> grid.rect(x=0.66, height=0.7, width=0.2,
            gp=gpar(fill="black"))
> popViewport()
```

The graphical parameter settings in a viewport only affect other viewports and graphical output within that viewport. The settings do not affect the viewport itself. For example, parameters controlling the size of text (fontsize, cex, etc.) do not affect the meaning of "lines" units when determining the location and size of the viewport (but they will affect the location and size of other viewports or graphical output within the viewport). A layout (see Section 5.5.6) counts as being within the viewport (i.e., it is affected by the graphical parameter settings of the viewport).

If there are multiple values for a graphical parameter setting, only the first is used when determining the location and size of a viewport.

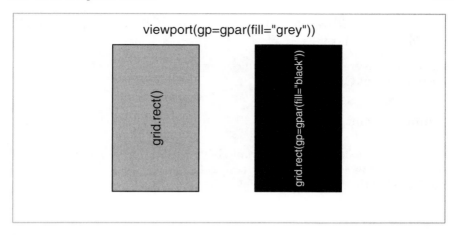

Figure 5.15
The inheritance of viewport graphical parameters. A diagram demonstrating how viewport graphical parameter settings are inherited by graphical output within the viewport. The viewport sets the default fill color to grey. The left-hand rectangle specifies no fill color itself so it is filled with grey. The right-hand rectangle specifies a black fill color that overrides the viewport setting.

5.5.6 Layouts

A viewport can have a *layout* specified via the `layout` argument. A layout in grid is similar to the same concept in traditional graphics (see Section 3.3.2). It divides the viewport region into several columns and rows, where each column can have a different width and each row can have a different height. For several reasons, however, layouts are much more flexible in grid: there are many more coordinate systems for specifying the widths of columns and the heights of rows (see Section 5.3); viewports can occupy overlapping areas within the layout; and each viewport within the viewport tree can have a layout (layouts can be nested). There is also a `just` argument to justify the layout within a viewport when the layout does not occupy the entire viewport region.

Layouts provide a convenient way to position viewports using the standard set of coordinate systems, and provide an extra coordinate system, `"null"`, which is specific to layouts.

The basic idea is that a viewport can be created with a layout and then subsequent viewports can be positioned relative to that layout. In simple cases, this can be just a convenient way to position viewports in a regular grid, but in more complex cases, layouts are the only way to apportion regions. There are very many ways that layouts can be used in grid; the following

sections attempt to provide a glimpse of the possibilities by demonstrating a
series of example uses.

A grid layout is created using the function `grid.layout()` (*not* the traditional
function `layout()`).

A simple layout

The following code produces a simple layout with three columns and three
rows, where the central cell (row two, column two) is forced to always be
square (using the `respect` argument).

```
> vplay <- grid.layout(3, 3,
                       respect=rbind(c(0, 0, 0),
                                     c(0, 1, 0),
                                     c(0, 0, 0)))
```

The next piece of code uses this layout in a viewport. Any subsequent viewports may make use of the layout, or they can ignore it completely.

```
> pushViewport(viewport(layout=vplay))
```

In the next piece of code, two further viewports are pushed within the viewport
with the layout. The `layout.pos.col` and `layout.pos.row` arguments are
used to specify which cells within the layout each viewport should occupy. The
first viewport occupies all of column two and the second viewport occupies all
of row 2. This demonstrates that viewports can occupy overlapping regions
within a layout. A rectangle has been drawn within each viewport to show
the region that the viewport occupies (see Figure 5.16).

```
> pushViewport(viewport(layout.pos.col=2, name="col2"))
> upViewport()
> pushViewport(viewport(layout.pos.row=2, name="row2"))
```

A layout with units

This section describes a layout that makes use of grid units. In the context of
specifying the widths of columns and the heights of rows for a layout, there is
an additional unit available, the `"null"` unit. All other units (`"cm"`, `"npc"`,
etc.) are allocated first within a layout, then the `"null"` units are used to
divide the remaining space proportionally (see Section 3.3.2). The following

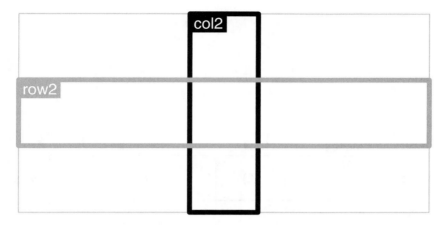

Figure 5.16
Layouts and viewports. Two viewports occupying overlapping regions within a layout. Each viewport is represented by a rectangle with the viewport name at the top-left corner. The layout has three columns and three rows with one viewport occupying all of row 2 and the other viewport occupying all of column 2.

code creates a layout with three columns and three rows. The left column is one inch wide and the top row is three lines of text high. The remainder of the current region is divided into two rows of equal height and two columns with the right column twice as wide as the left column (see Figure 5.17).

```
> unitlay <-
    grid.layout(3, 3,
                widths=unit(c(1, 1, 2),
                            c("inches", "null", "null")),
                heights=unit(c(3, 1, 1),
                            c("lines", "null", "null")))
```

With the use of `"strwidth"` and `"grobwidth"` units it is possible to produce columns that are just wide enough to fit graphical output that will be drawn in the column (and similarly for row heights — see Section 6.4).

A nested layout

This section demonstrates the nesting of layouts. The following code defines a function that includes a trivial use of a layout consisting of two equal-width columns to produce grid output.

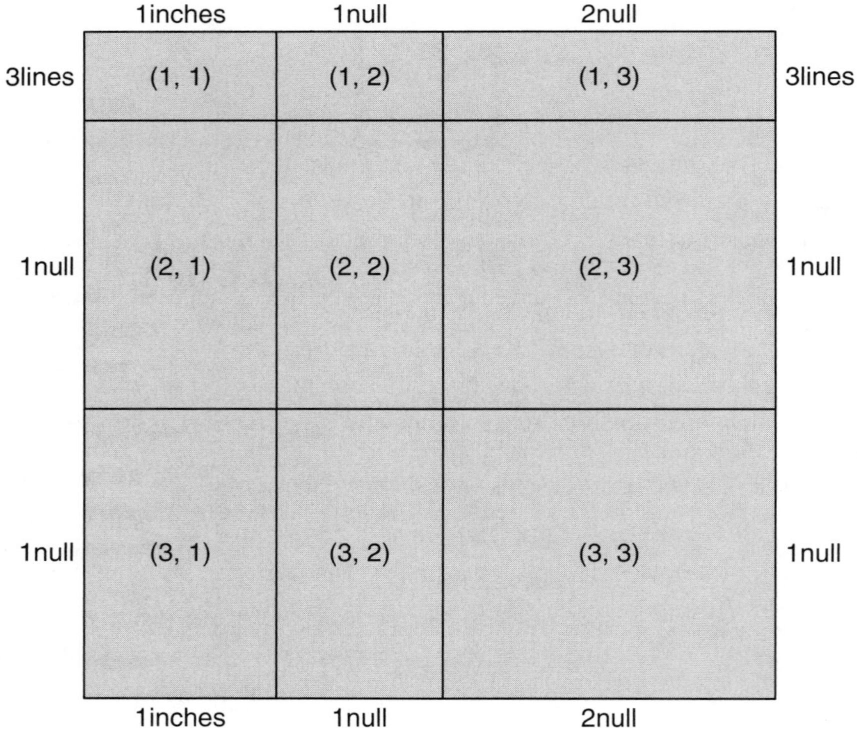

	1inches	1null	2null	
3lines	(1, 1)	(1, 2)	(1, 3)	3lines
1null	(2, 1)	(2, 2)	(2, 3)	1null
1null	(3, 1)	(3, 2)	(3, 3)	1null
	1inches	1null	2null	

Figure 5.17

Layouts and units. A grid layout using a variety of coordinate systems to specify the widths of columns and the heights of rows.

```
> gridfun <- function() {
    pushViewport(viewport(layout=grid.layout(1, 2)))
    pushViewport(viewport(layout.pos.col=1))
    grid.rect()
    grid.text("black")
    grid.text("&", x=1)
    popViewport()
    pushViewport(viewport(layout.pos.col=2, clip="on"))
    grid.rect(gp=gpar(fill="black"))
    grid.text("white", gp=gpar(col="white"))
    grid.text("&", x=0, gp=gpar(col="white"))
    popViewport(2)
  }
```

The next piece of code creates a viewport with a layout and places the output
from the above function within a particular cell of that layout (see Figure
5.18).

```
> pushViewport(
    viewport(
      layout=grid.layout(5, 5,
                         widths=unit(c(5, 1, 5, 2, 5),
                                     c("mm", "null", "mm",
                                       "null", "mm")),
                         heights=unit(c(5, 1, 5, 2, 5),
                                      c("mm", "null", "mm",
                                        "null", "mm")))))
> pushViewport(viewport(layout.pos.col=2, layout.pos.row=2))
> gridfun()
> popViewport()
> pushViewport(viewport(layout.pos.col=4, layout.pos.row=4))
> gridfun()
> popViewport(2)
```

Although the result of this particular example could be achieved using a single
layout, what this shows is that it is possible to take grid code that makes use
of a layout (and may have been written by someone else) and embed it within
a layout of your own. A more sophisticated example of this involving lattice
plots is given in Section 5.8.2.

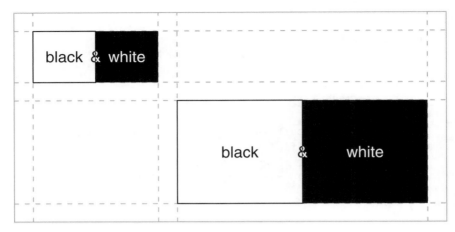

Figure 5.18

Nested layouts. An example of a layout nested within a layout. The black and white
squares are drawn within a layout that has two equal-width columns. One instance
of the black and white squares has been embedded within cell $(2, 2)$ of a layout
consisting of five columns and five rows of varying widths and heights (as indicated
by the dashed lines). Another instance has been embedded within cell $(4, 4)$.

5.6   Missing values and non-finite values

Non-finite values are not permitted in the location, size, or scales of a viewport.
Viewport scales are checked when a viewport is created, but it is impossible
to be certain that locations and sizes are not non-finite when the viewport
is created, so this is only checked when the viewport is pushed. Non-finite
values result in error messages.

The locations and sizes of graphical objects can be specified as missing values
(NA, "NA") or non-finite values (NaN, Inf, -Inf). For most graphical primitives,
non-finite values for locations or sizes result in the corresponding primitive
not being drawn. For the grid.line.to() function, a line segment is only
drawn if the previous location and the new location are both not non-finite.
For grid.polygon(), a non-finite value breaks the polygon into two separate
polygons. This break happens within the current polygon as specified by the
id argument. All polygons with the same id receive the same gp settings. For
grid.arrows(), an arrow head is only drawn if the first or last line segment
is drawn.

Figure 5.19 shows the behavior of these primitives where x- and y-locations are seven equally-spaced locations around the perimeter of a circle. In the top-left figure, all locations are not non-finite. In each of the other figures, two locations have been made non-finite (indicated in each case by grey text).

5.7 Interactive graphics

The strength of the grid system is in the production of static graphics. There is only very basic support for user interaction, consisting of the `grid.locator()` function. This function returns the location of a single mouse click relative to the current viewport. The result is a list containing an `x` and a `y` unit. The `unit` argument can be used to specify the coordinate system to be used for the result.

From R version 2.1.0, the `getGraphicsEvent()` function provides additional capability (on Windows) to respond to mouse movements, mouse ups, and key strokes. However, with this function, mouse activity is only reported relative to the native coordinate system of the device.

5.8 Customizing lattice plots

This section provides some demonstrations of the basic grid functions within the context of a complete lattice plot.

The lattice package described in Chapter 4 produces complete and very sophisticated plots using grid. It makes use of a sometimes large number of viewports to arrange the graphical output. A page of lattice output contains a top-level viewport with a quite complex layout that provides space for all of the panels and strips and margins used in the plot. Viewports are created for each panel and for each strip (among other things), and the plot is constructed from a large number of rectangles, lines, text, and data points.

In many cases, it is possible to use lattice without having to know anything about grid. However, a knowledge of grid provides a number of more advanced ways to work with lattice output (see Section 6.7). A simple example is provided by the `panel.width` and `panel.height` arguments to the `print.trellis()` method. These provide an alternative to the `aspect` argument for controlling the size of panels within a lattice plot using grid units.

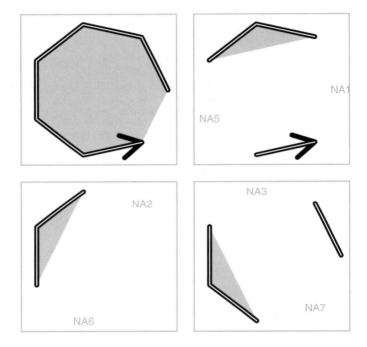

Figure 5.19
Non-finite values for line-tos, polygons, and arrows. The effect of non-finite values
for `grid.line.to()`, `grid.polygon()`, and **grid.arrows**. In each panel, a single
grey polygon, a single arrow (at the end of a thick black line), and a series of thin
white line-tos are drawn through the same set of seven points. In some cases, certain
locations have been set to **NA** (indicated by grey text), which causes the polygon to
become cropped, creates gaps in the lines, and can cause the arrow head to disappear.
In the bottom-left panel, the seventh location is not **NA**, but it produces no output.

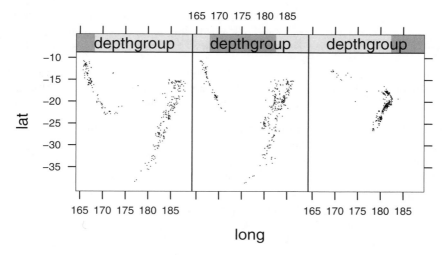

Figure 5.20
Controlling the size of lattice panels using grid units. Each panel is exactly 1.21 inches wide and 1.5 inches high.

The following code produces a multipanel lattice plot of the quakes data set (see page 126) where the size of each panel is fixed at 1.21 inches wide and 1.5 inches high (see Figure 5.20).*

```
> temp <- xyplot(lat ~ long | depthgroup,
                 data=quakes, pch=".",
                 layout=c(3, 1))
> print(temp,
        panel.width=list(1.21, "inches"),
        panel.height=list(1.5, "inches"))
```

5.8.1   Adding grid output to lattice output

The functions that lattice provides for adding output to panels (ltext(), lpoints(), etc) are designed to make it easier to port code between R and S-PLUS. However, they are restricted because they only allow output to be located and sized relative to the "native" coordinate system. Grid graphical primitives cannot be ported to S-PLUS, but they provide much more control

---
*These specific sizes were chosen for this particular data set so that one unit of longitude corresponds to the same physical size on the page as one unit of latitude.

over the location and size of additional panel output. Furthermore, it is possible to create and push extra viewports within a panel if desired (although it is very important that they are popped again or lattice will get very confused).

In a similar vein, the facilities provided by the `upViewport()` and `downViewport()` functions in grid allow for more flexible navigation of a lattice plot compared to the `trellis.focus()` function.

The following code provides an example of using low-level grid functions to add output within a lattice panel function. This produces a variation on Figure 4.4 with a dot and a text label added to indicate the location of Auckland, New Zealand relative to the earthquakes (see Figure 5.21).*

```
> xyplot(lat ~ long | depthgroup, data=quakes, pch=".",
         panel=function(...) {
           grid.points(174.75, -36.87, pch=16,
                       size=unit(2, "mm"),
                       default.units="native")
           grid.text("Auckland",
                     unit(174.75, "native") - unit(2, "mm"),
                     unit(-36.87, "native"),
                     just="right")
           panel.xyplot(...)
         })
```

5.8.2 Adding lattice output to grid output

As well as the advantages of using grid functions to add further output to lattice plots, an understanding that lattice output is really grid output makes it possible to embed lattice output within grid output. The following code provides a simple example (see Figure 5.22).

First of all, two viewports are defined. The viewport `tvp` occupies the rightmost 1 inch of the device and will be used to draw a label. The viewport `lvp` occupies the rest of the device and will be used to draw a lattice plot.

```
> lvp <- viewport(x=0,
                  width=unit(1, "npc") - unit(1, "inches"),
                  just="left", name="lvp")
> tvp <- viewport(x=1, width=unit(1, "inches"),
                  just="right", name="tvp")
```

---

*The data are from the `quakes` data set (see page 126).

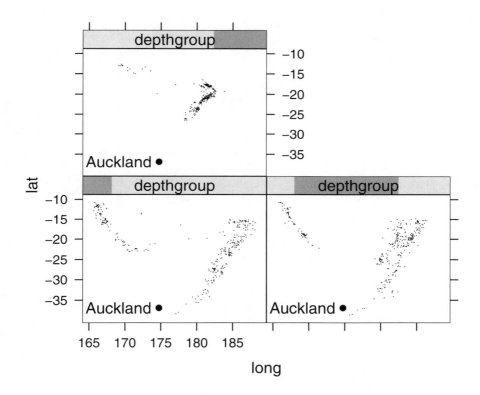

Figure 5.21
Adding grid output to a lattice plot (the lattice plot in Figure 4.4). The grid
functions grid.text() and grid.points() are used within a lattice panel function
to highlight the location of Auckland, New Zealand within each panel.

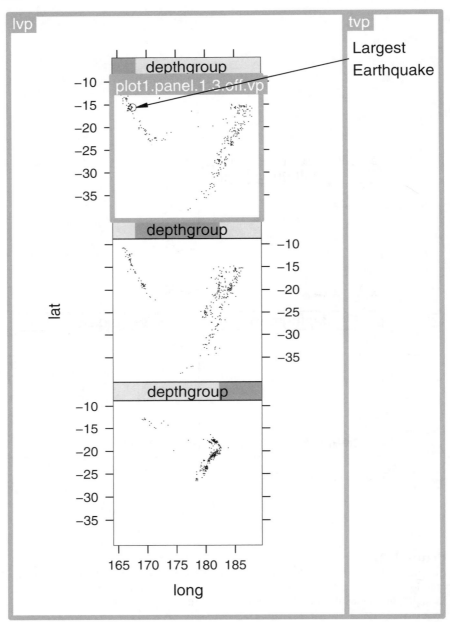

Figure 5.22
Embedding a lattice plot within grid output. The lattice plot is drawn within
the viewport `"lvp"` and the text label is drawn within the viewport `"tvp"` (the
viewports are indicated by grey rectangles with their names at the top-left corner).
An arrow is drawn from viewport `"tvp"` where the text was drawn into viewport
`"panel.1.3.off.vp"` — the top panel of the lattice plot.

The next piece of code produces (but does not draw) an object representing a multipanel scatterplot using the **quakes** data (see page 126).

```
> lplot <- xyplot(lat ~ long | depthgroup,
                  data=quakes, pch=".",
                  layout=c(1, 3), aspect=1,
                  index.cond=list(3:1))
```

The following pieces of code do all the drawing. First of all, the **lvp** viewport is pushed and the lattice plot is drawn inside that. The **upViewport()** function is used to navigate back up so that all of the lattice viewports are left intact.

```
> pushViewport(lvp)
> print(lplot, newpage=FALSE, prefix="plot1")
> upViewport()
```

Next, the **tvp** viewport is pushed and a text label is drawn in that.

```
> pushViewport(tvp)
> grid.text("Largest\nEarthquake", x=unit(2, "mm"),
            y=unit(1, "npc") - unit(0.5, "inches"),
            just="left")
```

The last step is to draw an arrow from the label to a data point within the lattice plot. While still in the **tvp** viewport, the **grid.move.to()** function is used to set the current location to a point just to the left of the text label. Next, **seekViewport()** is used to navigate to the top panel within the lattice plot.* Finally, **grid.arrows()** and **lineToGrob()** are used to draw a line from the text to an (x ,y) location within the top panel. A circle is also drawn to help identify the location being labelled.

---

*The name of the viewport representing the top panel in the lattice plot can be obtained using the **trellis.vpname()** function or by just visual inspection of the output of **current.vpTree()** and possibly some trial-and-error.

```
> grid.move.to(unit(1, "mm"),
                unit(1, "npc") - unit(0.5, "inches"))
> seekViewport("plot1.panel.1.3.off.vp")
> grid.arrows(grob=lineToGrob(unit(167.62, "native") +
                                unit(1, "mm"),
                                unit(-15.56, "native")),
                length=unit(3, "mm"), type="closed",
                angle=10, gp=gpar(fill="black"))
> grid.circle(unit(167.62, "native"),
                unit(-15.56, "native"),
                r=unit(1, "mm"),
                gp=gpar(lwd=0.1))
```

The final output is shown in Figure 5.22.

Chapter summary

Grid provides a number of functions for producing basic graphical output such as lines, polygons, rectangles, and text, plus some functions for producing slightly more complex output such as data symbols, arrows, and axes. Graphical output can be located and sized relative to a large number of coordinate systems and there are a number of graphical parameter settings for controlling the appearance of output, such as colors, fonts, and line types.

Viewports can be created to provide contexts for drawing. A viewport defines a rectangular region on the device and all coordinate systems are available within all viewports. Viewports can be arranged using layouts and nested within one another to produce sophisticated arrangements of graphical output.

Because lattice output is grid output, grid functions can be used to add further output to a lattice plot. Grid functions can also be used to control the size and placement of lattice plots.

6

The Grid Graphics Object Model

Chapter preview

This chapter describes how to work with graphical output as graphical objects (grobs). The main advantage of this approach is that it is possible to interactively edit a scene that was produced using grid. Because lattice is built on grid, this means it is possible to interactively edit a lattice plot.

There are also benefits from being able to do such things as ask a piece of graphical output how big it is. For example, this makes it easy to leave space for a legend beside a plot.

Graphical objects can be combined to form larger, hierarchical graphical objects (gTrees). This makes it possible to control the appearance and position of whole groups of graphical objects at once.

This chapter describes the grid concepts of grobs and gTrees as well as important functions for accessing, querying, and modifying these objects.

The previous chapter mostly dealt with using grid functions to produce graphical output. That knowledge is useful for annotating a plot produced using grid (such as a lattice plot), for producing one-off or customized plots for your own use, and for writing simple graphics functions.

This chapter on the other hand addresses grid functions for creating and manipulating graphical objects. This information is useful for interactively editing or modifying graphical output and for writing graphical functions and objects for others to use (also see Chapter 7).

6.1   Working with graphical output

This section describes using grid to interactively modify graphical output. Having called functions to draw some output, there are functions to edit and delete elements of the output.

All of the functions in the previous section that produce graphical output also produce graphical objects, or *grobs*, representing that output. For example, the following code produces a number of circles as output (see the left panel in Figure 6.1).

```
> grid.circle(name="circles", x=seq(0.1, 0.9, length=40),
              y=0.5 + 0.4*sin(seq(0, 2*pi, length=40)),
              r=abs(0.1*cos(seq(0, 2*pi, length=40))))
```

As well as drawing the circles, this code produces a `circle` grob, an object of class `"circle"`, which contains information describing the circles that have been drawn.

Grid maintains a display list, a record of all viewports and grobs drawn on the current page, and the object that `grid.circle()` created is stored on this display list. This means that it can be accessed to obtain a copy, to modify the output, or even to remove it altogether. The grob has been given the name `"circles"` to make it easy to identify.

In the following code, the call to `grid.get()` obtains a copy of the `circle` object. This can be useful for querying the elements of a scene (e.g., to see what components an element has).

```
> grid.get("circles")
```

```
circle[circles]
```

The following call to `grid.edit()` modifies the output by editing the `circle` object to change the colors used for drawing the circles (see the middle panel of Figure 6.1). In this case, the `gp` component of the `circle` grob is being modified. Typically, most arguments that can be specified when first drawing output can also be used when editing output.

```
> grid.edit("circles",
            gp=gpar(col=grey(c(1:20*0.04, 20:1*0.04))))
```

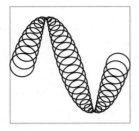 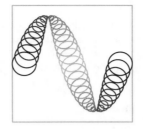 

Figure 6.1
Modifying a `circle` grob. The left panel shows the output produced by a call to
`grid.circle()`, the middle panel shows the result of using `grid.edit()` to modify
the colors of the circles, and the right panel shows the result of using `grid.remove()`
to delete the circles.

The next call below, to the `grid.remove()` function, deletes the output by
removing the `circle` object from the display list (see the right panel of Figure
6.1).

```
> grid.remove("circles")
```

In each of these examples, the grob has been specified by giving its name
(`"circles"`). Other standard arguments to these functions are discussed in
the next section.

Any output produced by grid functions can be interacted with in this way,[*]
including output from lattice functions (see Section 6.7).

6.1.1 Standard functions and arguments

The complete set of functions that provide the ability to interact with grobs
is given in Table 6.1.

All of the functions for working with graphical output require a grob name as
the first argument, to identify which grob to work with. This name will be
treated as a regular expression if the `grep` argument is `TRUE`.

If the `global` argument is `TRUE` then all matching grobs on the display list
(not just the first) will be accessed or modified.

The following code provides a simple example. Eight concentric `circle` grobs
are drawn, with the first, third, fifth, and seventh circles named `"circle.odd"`

---

[*] It is possible to disable the grid display list, in which case no grobs are stored so these
sorts of manipulations are no longer possible.

Table 6.1

Functions for working with grobs. Functions of the form `grid.*()` access and destructively modify grobs on the grid display list and affect graphical output. Functions of the form `*Grob()` work with user-level grobs and return grobs as their values (they have no effect on graphical output).

Function to work with output	Description	Function to work with grobs
`grid.get()`	Returns a copy of one or more grobs	`getGrob()`
`grid.edit()`	Modifies one or more grobs	`editGrob()`
`grid.add()`	Adds a grob to one or more grobs	`addGrob()`
`grid.remove()`	Removes one or more grobs	`removeGrob()`
`grid.set()`	Replaces one or more grobs	`setGrob()`

and the second, fourth, sixth, and eighth circles named `"circle.even"`. The circles are initially drawn with decreasing shades of grey (see the left panel of Figure 6.2).

```
> suffix <- c("even", "odd")
> for (i in 1:8)
     grid.circle(name=paste("circle.", suffix[i %% 2 + 1],
                            sep=""),
                 r=(9 - i)/20,
                 gp=gpar(col=NA, fill=grey(i/10)))
```

The following call to `grid.edit()` makes use of the `global` argument to modify all grobs named `"circle.odd"` and change their fill color to a very dark grey (see the middle panel of Figure 6.2).

```
> grid.edit("circle.odd", gp=gpar(fill="grey10"),
           global=TRUE)
```

A second call to `grid.edit()`, below, makes use of both the `grep` argument and the `global` argument to modify all grobs with names matching the pattern `"circle"` (all of the circles) and change their fill color to a light grey and their border color to a darker grey (see the right panel of Figure 6.2).

```
> grid.edit("circle", gp=gpar(col="grey", fill="grey90"),
           grep=TRUE, global=TRUE)
```

Figure 6.2

Editing grobs using `grep` and `global` in `grid.edit()`. The left-hand panel shows eight separate concentric circles, with names alternating between `"circle.odd"` and `"circle.even"`, filled with progressively lighter shades of grey. The middle panel shows the use of the `global` argument to change the fill for all circles named `"circle.odd"` to black. The right-hand panel shows the use of the `grep` and `global` arguments to change all circles whose names match the pattern `"circle"` (all of the circles) to have a light grey fill and a grey border.

In summary, as long as the name of a grob is known, it is possible to access that grob using `grid.get()`, modify it using `grid.edit()`, or delete it using `grid.remove()`.

The function `getNames()` is useful for producing a list of all grobs in the current scene.*

6.2   Grob lists, trees, and paths

As well as basic grobs, it is possible to work with a list of grobs (a gList) or several grobs combined together in a tree-like structure (a gTree). A gList is just a list of several grobs (produced by the function `gList()`). A gTree is a grob that can contain other grobs. Examples are the `xaxis` and `yaxis` grobs. This section looks at how to work with gTrees.

An `xaxis` grob contains a high-level description of an axis, plus several child grobs representing the lines and text that make up the axis (see Figure 6.3).

The following code draws an xaxis and creates an `xaxis` grob on the display list (see the left panel of Figure 6.4). The `grid.get()` function is used to

---

*The `getNames()` function was only introduced in R version 2.1.0; in R 2.0.0, the expression `grid.get(".*", grep=TRUE, global=TRUE)` does something similar, but is less efficient.

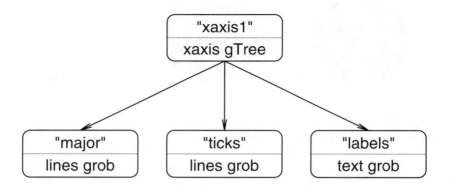

Figure 6.3
The structure of a gTree. A diagram of the structure of an xaxis gTree. There is
the xaxis gTree itself (here given the name "xaxis1") and there are its children: a
lines grob named "major", another lines grob named "ticks", and a text grob
named "labels".

obtain a copy of the grob and the childNames() function returns the names
of the grobs that are children of the xaxis grob.

```
> grid.xaxis(name="axis1", at=1:4/5)
> childNames(grid.get("axis1"))

[1] "major"  "ticks"  "labels"
```

The hierarchical structure of gTrees makes it possible to interact with both a
high-level description, as provided by the xaxis grob, and a low-level description, as provided by the children of the gTree. The following code demonstrates
an interaction with the high-level description of an xaxis grob. The xaxis
gTree contains components describing where to put tick-marks on the axis and
whether to draw labels and so on. The code below shows the at component of
an xaxis grob being modified. The xaxis grob is designed so that it modifies
its children to match the new high-level description so that only three ticks
are now drawn (see the middle panel of Figure 6.4).

```
> grid.edit("axis1", at=1:3/4)
```

It is also possible to access the children of a gTree. In the case of an xaxis,
there are three children: a lines grob with the name "major"; another lines

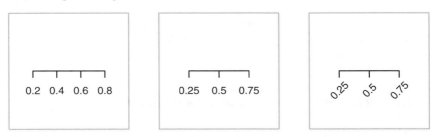

Figure 6.4
Editing a gTree. The left-hand panel shows a basic x-axis, the middle panel shows
the effect of editing the **at** component of the x-axis (all of the tick marks and
labels have changed), and the right-hand panel shows the effect of editing the **rot**
component of the **"labels"** child of the x-axis (only the angle of rotation of the
labels has changed).

grob with the name **"ticks"**; and a **text** grob with the name **"labels"**. Any
of these children can be accessed by specifying the name of the **xaxis** grob
and the name of the child in a grob path (gPath). A gPath is like a viewport
path (see Section 5.5.3) — it is just a concatenation of several grob names.
The following code shows how to access the **"labels"** child of the **xaxis** grob
using the **gPath()** function to specify a gPath. The gPath specifies the child
called **"label"** in the gTree called **"axis1"**. The labels are rotated to 45
degrees (see the right panel of Figure 6.4).

```
> grid.edit(gPath("axis1", "labels"), rot=45)
```

It is also possible to specify a gPath directly as a string, for example
"axis1::labels", but this is only recommended for interactive use.

6.2.1   Graphical parameter settings in gTrees

Just like any other grob, a gTree can have graphical parameter settings associated with it via a **gp** component. These settings affect all graphical objects
that are children of the gTree, unless the children specify their own graphical parameter setting. In other words, the graphical parameter settings for a
gTree modify the implicit graphical context for the children of the gTree (see
page 168).

The following expression demonstrates this rule. The **gp** component of an
xaxis grob sets the drawing color to be **"grey"**. This means that all of the
children of the **xaxis** — the lines and labels — will be drawn grey.

```
> grid.xaxis(gp=gpar(col="grey"))
```

Another example of this behavior is given in Section 6.3 and the role of the
gp component in the drawing behavior of gTree objects is described in more
detail in Section 7.3.4.

6.2.2   Viewports as components of gTrees

Just like any other grob, a gTree can have a viewport (or viewport tree, or
viewport path, etc.) associated with it via a **vp** component. This viewport is
pushed before the gTree is drawn and popped afterwards (see Section 5.5.4).
This means that the children of a gTree are drawn within the drawing context
defined by the viewport in the **vp** slot of the gTree (see page 173).

The following code demonstrates this rule. The **vp** component of an **xaxis**
grob specifies a viewport in the top half of the page. This means that the
children of the **xaxis** are positioned relative to that viewport.

```
> grid.xaxis(vp=viewport(y=0.75, height=0.5))
```

An example of this behavior is given in Section 6.3 and the role of the **vp**
component in the drawing behavior of gTree objects is described in more
detail in Sections 7.3.4 and 7.3.7.

6.2.3   Searching for grobs

This section provides details about how grob names and gPaths are used to
find a grob.

Grobs are stored on the grid display list in the order that they are drawn.
When searching for a matching name, the functions in Table 6.1 search the
display list from the beginning. This means that if there are several grobs
whose names are matched, they will be found in the order that they were
drawn.

Furthermore, the functions perform a depth-first search. This means that if
there is a gTree on the display list, and its name is not matched, then its
children are searched for a match before any other grobs on the display list
are searched.

The name to search for can be given as a gPath, which can be useful
to explicitly specify a particular child grob of a particular gTree (as in
"axis1::labels").

The argument **strict** controls whether a complete match must be found.
By default, the **strict** argument is **FALSE**, so in the previous exam-

ple, the "labels" child of "axis1" could have been accessed simply by
grid.get("labels"). On the other hand, if strict is set to TRUE, then
simply specifying "labels" results in no match as shown by the following
code (there is no top-level grob with the name "labels").

```
> grid.edit("labels", strict=TRUE, rot=45)
```

Error in
  editDLfromGPath(gPath, specs, strict, grep, global, redraw) :

    'gPath' (labels) not found

6.3 Working with graphical objects off-screen

The previous section described how grid functions that produce graphical
output also produce grobs. In some cases, it is useful to create a grob without
producing any output. This section describes how to use grid to produce
graphical objects (without drawing them). There are functions to create grobs,
functions to combine them, and modify them, and the grid.draw() function
to draw them.

For each grid function that produces graphical output, there is a counterpart
that produces a graphical object and no graphical output. For example, the
counterpart to grid.circle() is the function circleGrob() (see Table 5.1).
Similarly, for each function that works with grobs on the grid display list,
there is a counterpart for working with grobs off-screen. For example, the
counterpart to grid.edit() is editGrob() (see Table 6.1).

The following example demonstrates the process of creating a grob and working with a grob without drawing it. The code below draws a rectangle that is
as wide as a text grob, but the text is not drawn. The function textGrob()
produces a text grob, but does not draw it.

```
> grid.rect(width=grobWidth(textGrob("Some text")))
```

It can be useful to create a grob and modify it before producing any graphical
output (i.e., only draw the final result). The following code creates an axis
and modifies the font face for the labels to italic before drawing the axis. The
function grid.draw() is used to produce graphical output from a grob.

```
> ag <- xaxisGrob(at=1:4/5)
> ag <- editGrob(ag, "labels", gp=gpar(fontface="italic"))
> grid.draw(ag)
```

Another example of working with grobs is in the construction of gTrees. In its simplest form, a gTree is just a grouping of several grobs (more complex gTree creation is discussed later in Section 7.3).

By grouping several grobs together as a single object, the grobs can be dealt with as a single object. For example, it becomes possible to edit the graphical context for all of the grobs at once, or define the drawing context for all of the grobs at once.

When a gTree is drawn, any viewports in its vp component are pushed, any settings in its gp component are enforced, and then its children are all drawn. This means that the vp and gp components of a gTree affect where and how the children of the gTree are drawn (see Sections 6.2.1 and 6.2.2).

As an example, a boxed-text object can be created by grouping a "rect" grob and a "text" grob together as children of a gTree. This allows us to modify the color of both the rectangle and the text by modifying these features in the gTree. Similarly, it is possible to locate both the rectangle and the text by defining a viewport for the gTree.

The following code uses the gTree() function to create a gTree that groups a "rect" grob and a "text" grob together. There is no graphical output produced from this code. It only creates graphical objects.

```
> tg <- textGrob("sample text")
> rg <- rectGrob(width=1.1*grobWidth(tg),
                 height=1.3*grobHeight(tg))
> boxedText <- gTree(children=gList(tg, rg))
```

It is now easy to produce output including both the rectangle and the text by drawing variations on the boxedText grob, as demonstrated by the following code.

The first call simply draws the plain boxedText, which is drawn in black (see the left panel of Figure 6.5).

```
> grid.draw(boxedText)
```

The second call draws a modified boxedText with the drawing color set to grey (see the middle panel of Figure 6.5).

```
> grid.draw(editGrob(boxedText, gp=gpar(col="grey")))
```

Figure 6.5
Using a gTree to group grobs. The left-hand panel shows a boxed text object (which is a combination of a piece of text and a rectangle). The middle panel shows how changes to the color settings in the boxed text object propagate to the components (both the text and rectangle turn grey). The right-hand panel shows a more dramatic demonstration of the same idea as, in this case, the fontsize of the boxed text is modified and it is drawn within a rotated viewport.

The final call draws another modification of the `boxedText`, this time in a rotated viewport and with a larger font (see the right panel of Figure 6.5).

```
> grid.draw(editGrob(boxedText, vp=viewport(angle=45),
                     gp=gpar(fontsize=18)))
```

6.3.1 Capturing output

In the example in the previous section, several grobs are created off-screen and then grouped together as a gTree, which allows the collection of grobs to be dealt with as a single object.

It is also possible first to *draw* several grobs and *then* to group them. The `grid.grab()` function does this by generating a gTree from all of the grobs in the current page of output. This means that output can be captured even from a function that produces very complex output (lots of grobs), such as a lattice plot. For example, the following code draws a lattice plot, then creates a gTree containing all of the grobs in the plot.

```
> bwplot(rnorm(10))
> bwplotTree <- grid.grab()
```

The `grid.grab()` function actually captures all of the viewports in the current scene as well as the grobs, so drawing the gTree, as in the following code, produces exactly the same output as the original plot.

```
> grid.newpage()
> grid.draw(bwplotTree)
```

Another function, `grid.grabExpr()` allows complex output to be captured off-screen. This function takes an R expression and evaluates it. Any drawing that occurs as a result of evaluating the expression does not produce any output, but the grobs that would be produced are captured anyway.

The following code provides a simple demonstration. Here a lattice plot is captured without drawing any output.*

```
> grid.grabExpr(print(bwplot(rnorm(10))))
```

gTree[GRID.GROB.152]

Both the `grid.grab()` and `grid.grabExpr()` functions attempt to create a gTree in a sophisticated way so that it is easier to work with the resulting gTree. Unfortunately, this will not always produce a gTree that will exactly replicate the original output. These functions issue warnings if they detect a situation where output may not be reproduced correctly, and there is a `wrap` argument that can be used to force the functions to produce a gTree that is less sophisticated, but is guaranteed to replicate the original output.†

6.4   Placing and packing grobs in frames

It can be useful to position the components of a plot in a way that leaves sufficient room for labels or legends. The `"grobwidth"` and `"grobheight"` coordinate systems provide a way to determine the size of a grob and can be used to achieve this sort of arrangement of components by, for example, allocating appropriate regions within a layout.

The following code demonstrates this idea. First of all, some grobs are created to use as components of a scene. The first grob, `label`, is a simple `text` grob. The second grob, `gplot`, is a gTree containing a `rect` grob, a `lines` grob, and a `points` grob that provide a simple representation of time-series data. The

---

*The expression must explicitly `print()` the lattice plot because otherwise nothing would be drawn (see Section 4.1).

†The `grid.grabExpr()` function is only available from R version 2.1.0.

gplot has a viewport in its vp component and the rectangle and lines are drawn within that viewport.

```
> label <- textGrob("A\nPlot\nLabel ",
                    x=0, just="left")
> x <- seq(0.1, 0.9, length=50)
> y <- runif(50, 0.1, 0.9)
> gplot <-
    gTree(
      children=gList(rectGrob(gp=gpar(col="grey60",
                                      fill="white")),
                   linesGrob(x, y),
                   pointsGrob(x, y, pch=16,
                              size=unit(1.5, "mm"))),
      vp=viewport(width=unit(1, "npc") - unit(5, "mm"),
                  height=unit(1, "npc") - unit(5, "mm")))
```

The next piece of code defines a layout with two columns. The second column of the layout has its width determined by the width of the label grob created above. The first column will take up whatever space is left over.

```
> layout <- grid.layout(1, 2,
                        widths=unit(c(1, 1),
                                    c("null", "grobwidth"),
                                    list(NULL, label)))
```

Now some drawing can occur. A viewport is pushed with the layout defined above, then the label grob is drawn in the second column of this layout, which is exactly the right width to contain the text, and the gplot gTree is drawn in the first column (see Figure 6.6).

```
> pushViewport(viewport(layout=layout))
> pushViewport(viewport(layout.pos.col=2))
> grid.draw(label)
> popViewport()
> pushViewport(viewport(layout.pos.col=1))
> grid.draw(gplot)
> popViewport(2)
```

Grid provides a set of functions that make it more convenient to arrange grobs like this so that they allow space for each other. The function grid.frame(), and its off-screen counterpart frameGrob(), produce a gTree with no children. Children are added to the frame using the grid.pack() function and the

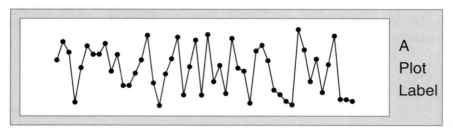

Figure 6.6
Packing grobs by hand. The scene was created using a frame object, into which the
time-series plot (consisting of a rectangle, lines, and points) was packed. The text
was then packed on the right-hand side, which meant that the time series plot was
allocated less room in order to leave space for the text.

frame makes sure that enough space is allowed for the child when it is drawn.
Using these functions, the previous example becomes simpler, as shown by the
following code (the output is the same as Figure 6.6). The big difference is
that there is no need to specify a layout as an appropriate layout is calculated
automatically.

The first call creates an empty frame. The second call packs `gplot` into the
frame; at this stage, `gplot` takes up the entire frame. The third call packs
the text label on the right-hand side of the frame; enough space is made for
the text label by reducing the space allowed for the rectangle.

```
> grid.frame(name="frame1")
> grid.pack("frame1", gplot)
> grid.pack("frame1", label, side="right")
```

There are many arguments to `grid.pack()` for specifying where to pack new
grobs within a frame. There is also a `dynamic` argument to specify whether
the frame should reallocate space if the grobs that have been packed in the
frame are modified.

Unfortunately, packing grobs into a frame like this becomes quite slow as more
grobs are packed, so it is most useful for very simple arrangements of grobs
or for interactively constructing a scene. An alternative approach, which
is a little more work, but still more convenient than dealing directly with
pushing and popping viewports (and can be made dynamic like packing), is
to *place* grobs within a frame that has a predefined layout. The following code
demonstrates this approach. This time, the frame is initially created with the
desired layout as defined above, then the `grid.place()` function is used to
position grobs within specific cells of the frame layout.

```
> grid.frame(name="frame1", layout=layout)
> grid.place("frame1", gplot, col=1)
> grid.place("frame1", label, col=2)
```

6.4.1 Placing and packing off-screen

In the previous two examples, the screen is redrawn each time a grob is packed into the frame. It is more typical to create a frame and pack or place grobs within it off-screen and only draw the frame once it is complete. The following code demonstrates the use of the `frameGrob()` and `placeGrob()` functions to achieve the same end result as shown in Figure 6.6, doing all of the construction of the frame off-screen.

```
> fg <- frameGrob(layout=layout)
> fg <- placeGrob(fg, gplot, col=1)
> fg <- placeGrob(fg, label, col=2)
> grid.draw(fg)
```

The function `packGrob()` is the off-screen counterpart of `grid.pack()`.

6.5 Other details about grobs

This section describes some important extra details about the calculation of grob sizes and the editing of graphical contexts.

6.5.1 Calculating the sizes of grobs

As described in Section 5.3.2, the `"grobwidth"` and `"grobheight"` units provide a way to determine the size of a grob. This section provides some more details about the correct usage of these units.

The most important point is that the size of a grob is always calculated relative to the current geometric and graphical context. The following code demonstrates this point. First of all, a `text` grob and a `rect` grob are created, and the dimensions of the `rect` grob are based on the dimensions of the text.[*]

---

[*]The `rect` grob draws two rectangles: one thick and dark grey; one white and thin.

```
> tg1 <- textGrob("Sample")
> rg1 <- rectGrob(x=rep(0.5, 2),
                  width=1.1*grobWidth(tg1),
                  height=1.3*grobHeight(tg1),
                  gp=gpar(col=c("grey60", "white"),
                          lwd=c(3, 1)))
```

Next, these two grobs are drawn in three different settings. In the first setting, the rectangle and the text are drawn in the default geometric and graphical context and the rectangle bounds the text (see the left panel of Figure 6.7).

```
> grid.draw(tg1)
> grid.draw(rg1)
```

In the second setting, the grobs are both drawn within a viewport that has cex=2. Both the text and the rectangle are drawn bigger (the calculation of the "grobwidth" and "grobheight" units takes place in the same context as the drawing of the text grob; see the middle panel of Figure 6.7).

```
> pushViewport(viewport(gp=gpar(cex=2)))
> grid.draw(tg1)
> grid.draw(rg1)
> popViewport()
```

In the third setting, the text grob is drawn in a different context than the rectangle, so the rectangle's size is "wrong" (see the right panel of Figure 6.7).

```
> pushViewport(viewport(gp=gpar(cex=2)))
> grid.draw(tg1)
> popViewport()
> grid.draw(rg1)
```

A related issue arises with the use of grob *names* when creating a "grobwidth" or "grobheight" unit (see Section 5.3.2). The following code provides a simple example.

A text grob and two rect grobs are created, with the dimensions of both rectangles based upon the dimensions of the text. One rectangle, rg1, uses a copy of the text grob in the calls to grobWidth(), and grobHeight(). The other rectangle, rg2, just uses the name of the text grob, "tg1".

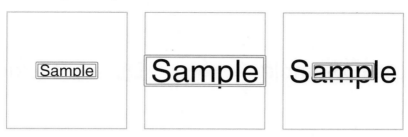

Figure 6.7
Calculating the size of a grob. In the left-hand panel, a `text` grob and a separate
`rect` grob, the size of which is calculated to be the size of the `text` grob, are drawn
together. In the middle panel, these objects are drawn together in a viewport with
a larger fontsize, so they are both larger. In the right-hand panel, only the text is
drawn in a viewport with a larger fontsize, so only the text is larger. The rectangle
calculates the size of the text in a different font context.

```
> tg1 <- textGrob("Sample", name="tg1")
> rg1 <- rectGrob(width=1.1*grobWidth("tg1"),
                  height=1.3*grobHeight("tg1"),
                  gp=gpar(col="grey60", lwd=3))
> rg2 <- rectGrob(width=1.1*grobWidth(tg1),
                  height=1.3*grobHeight(tg1),
                  gp=gpar(col="white"))
```

When these rectangles and text are initially drawn, both rectangles frame the
text correctly (see the left panel of Figure 6.8).

```
> grid.draw(tg1)
> grid.draw(rg1)
> grid.draw(rg2)
```

However, if the `text` grob is modified, as shown below, only the rectangle `rg1`
(the dark grey rectangle) will be updated to correspond to the new dimensions
of the text (see the right panel of Figure 6.8).

```
> grid.edit("tg1", grep=TRUE, global=TRUE,
            label="Different text")
```

With this approach, `"grobwidth"` and `"grobheight"` units are still evaluated
in the current geometric and graphical context, but in addition, only grobs
that have previously been drawn can be referred to. For example, drawing
the rectangle `rg1` before drawing the text `tg1` will not work because there is
no drawn grob named `"tg1"` from which a size can be calculated.

 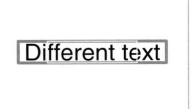

Figure 6.8
Grob dimensions by reference. In the left-hand panel there are three grobs: one
`text` grob and two `rect` grobs. The sizes of both `rect` grobs are calculated from the
`text` grob. The difference is that the white rectangle is related to the text by value
and the dark grey rectangle is related to the text by reference. The right-hand panel
shows what happens when the `text` grob is edited. Only the dark grey, by-reference,
rectangle gets resized.

```
> grid.newpage()
> grid.draw(rg1)
```

```
Error in function (name)  :
    Grob 'tg1' not found
```

6.5.2   Editing graphical context

When a grob is edited using `grid.edit()` or `editGrob()`, the modification of
a `gp` component is treated as a special case. Only the graphical parameters
that are explicitly given new settings are modified. All other settings remain
untouched. The following code provides a simple example.

A circle is drawn with a grey fill color (see the left panel of Figure 6.9), then
the border of the circle is made thick (see the middle panel of Figure 6.9) and
the fill color remains the same. Finally, the border is changed to a dashed
line type, but it stays thick (and the fill remains grey — see the right panel
of Figure 6.9).

```
> grid.circle(r=0.3, gp=gpar(fill="grey80"),
              name="mycircle")
> grid.edit("mycircle", gp=gpar(lwd=5))
> grid.edit("mycircle", gp=gpar(lty="dashed"))
```

 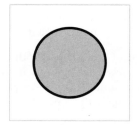 

Figure 6.9

Editing the graphical context. The left-hand panel shows a circle with a solid, thin black border and a grey fill. The middle panel shows the effect of making the border thicker. The important point is that the other features of the circle are not affected (the border is still solid and the fill is still grey). The right-hand panel shows another demonstration of the same idea, with the border now drawn dashed (but the border is still thick and the fill is still grey).

6.6 Saving and loading grid graphics

The best way to create a persistent record of a grid plot is to record in a text file the R code that was used to create the plot. The code can then be run again, e.g., using `source()`, to reproduce the output.

It is also possible to save grobs in R's binary format using the `save()` function. The grobs can then be loaded, using `load()`, and redrawn using `grid.draw()`. For the purpose of saving an entire scene, it may be more useful to save and load a gTree created by the `grid.grab` function (see Section 6.3.1).*

6.7 Working with lattice grobs

The output from a lattice function is fundamentally just a collection of grid viewports and grobs. Section 5.8 described some examples of interacting with

---

*A possible danger with saving a grid grob is that methods specific to that grob are not automatically recorded, so the grob may not behave correctly when loaded into a different session. This will only be an issue for grobs that are not predefined by the grid package (see Chapter 7, particularly Section 7.3).

the grid viewports that are set up by a lattice plot. This section looks at some examples of interacting with the grobs that are created by a lattice plot.

Unfortunately, only some of the grobs produced by lattice are given useful names. Many grobs only have the default names assigned them by the grid system, which are neither sufficiently descriptive nor reliable to be used to access a particular component of the plot.*

Examples of grobs that do have useful names are the `"xlab"` and `"ylab"` grobs representing the x-label and y-label on a lattice plot.

One thing that can be done with grobs produced by lattice is to modify them in ways that are not otherwise possible via the interface provided by lattice. For example, at the time of writing, it is possible to control the font face, the color, and the size of these labels (using `xlab` or `ylab` arguments to a plotting function or via `trellis.par.set()`), but it is not possible to modify the font family or the angle of rotation.

The following code produces a lattice scatterplot, then edits the plot labels to change the font to a `"mono"` family and to position the labels at the ends of the axes (see Figure 6.10).

```
> angle <- seq(0, 2*pi, length=21)[-21]
> x <- cos(angle)
> y <- sin(angle)

> xyplot(y ~ x, aspect=1,
          xlab="displacement",
          ylab="velocity")

> grid.edit("[.]xlab$", grep=TRUE,
            x=unit(1, "npc"), just="right",
            gp=gpar(fontfamily="mono"))
> grid.edit("[.]ylab$", grep=TRUE,
            y=unit(1, "npc"), just="right",
            gp=gpar(fontfamily="mono"))
```

Other grob operations are also possible. For example, the following code removes the labels from the plot.

```
> grid.remove(".lab$", grep=TRUE, global=TRUE)
```

---

*The default name assigned to a grob by the grid system is of the form `"GRID.GROB.`n`"` where n depends on how many other grobs have been created previously in the current session.

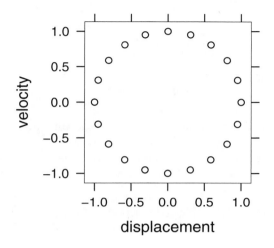

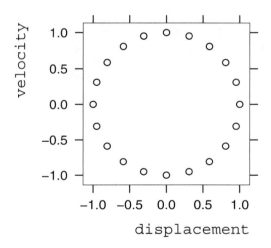

Figure 6.10
Editing the grobs in a lattice plot. The top plot is an initial scatterplot produced using the lattice function `xyplot()`. The bottom plot shows the effect of editing the grid `text` grobs that represent the labels on the plot (the labels are relocated at the ends of the axes and are drawn in a monospace font).

Finally, it is possible to group all of the grobs from a lattice plot together using `grid.grab()`. This creates a gTree that can then be used as a component in creating another picture.

Chapter summary

As well as producing graphical output, all grid functions create grobs (graphical objects) that contain descriptions of what has been drawn. These grobs may be accessed, modified, and even removed, and the graphical output will be updated to reflect the changes.

There are also grid functions for creating grobs without producing any graphical output. A complete description of a plot can be produced by creating, modifying, and combining grobs off-screen.

A gTree is a grob that can have other grobs as its children. A gTree can be useful for grouping grobs and for providing a high-level interface to a group of grobs.

The lattice plotting functions generate large numbers of grid grobs. These grobs may be manipulated just like any other grobs to access, edit, and delete parts of a lattice plot.

7

Developing New Graphics Functions and Objects

Chapter preview

This chapter looks in depth at the task of writing graphical functions for others to use.

There are important guidelines for writing simple functions whose main purpose is to produce graphical output. There is an emphasis on making sure that users can annotate the output produced by a function and that users can make use of the function as a component in larger or more complex plots.

There is also a discussion on how to create a new class of graphical object. This is important for allowing users to interactively edit output, to ask questions such as how much space a graphical object requires, and to be able to combine graphical objects together in a gTree.

This chapter addresses the issue of developing graphics functions for others to use. This will involve a discussion of some of the lower-level details of how grid works as well as some more abstract ideas of software design. A basic understanding of programming concepts is recommended, and the later sections assume an understanding of object-oriented concepts such as classes and methods.

Important low-level details of the grid graphics system and important design considerations are introduced in increasing levels of complexity to allow developers to construct simple graphics functions at first. Readers aiming to design a new fully-featured grid graphical object should read the entire chapter.

7.1   An example

In order to provide concrete examples of the concepts described in this chapter, a set of graphical functions and objects will be developed for the purpose of producing plots of oceanographic data.

An example of the final output that is desired is shown in Figure 7.1. Sections 7.2 to 7.3.8 go through the process of creating functions and objects to produce this output.

The data are measurements of fluorescence calculated at the thermocline (point of maximum temperature gradient) for 87 measuring stations off the coast of South Australia[38]. The values plotted in the image are from a prediction surface based on an analysis using the `Krig()` function in the `fields` package.[*]

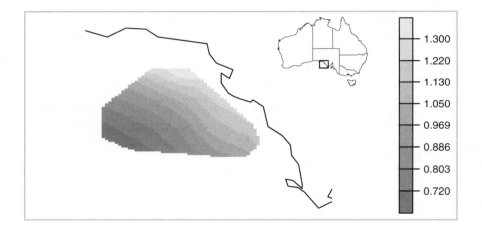

Figure 7.1
A plot of oceanographic data. The plot consists of a section of South Australian coastline, an image representing fluorescence at the thermocline, a small map to indicate where the main plot region is on Australia, and a legend to map the grey scale to fluorescence values.

---

[*]The data for the prediction surface are available as the data set `fluoro.predict` in the package `RGraphics`. Sam McClatchie provided the data and the original motivation to look at oceanographic plots in R.

7.1.1 Modularity

One decision can be made before writing a single line of code: the code should be modular. This means that the code should consist of several small functions, each of which produces a well-defined, self-contained piece of graphical output. It would be a bad idea to to create Figure 7.1 in one big function. Such a function would be unlikely to be very flexible, would be very hard to maintain (it is easier to see what is going on in smaller functions), and would be very hard to debug (it is much easier to test small functions with a clear, simple purpose).

The following sections look at writing several simple functions, each of which produces a conceptually separate part of the final plot. One possible breakdown of Figure 7.1 involves the following elements: two maps of Australia (one just a piece of the coastline), an array of colored rectangles (an *image*), and a legend. Immediately, the focus is on producing much more basic graphical output. If some useful functions are created for these, the functions will provide much more reusable graphical elements that could be combined in other ways to create all sorts of other plots (for example, see Section 7.3.9 and Figure 7.18).

7.2 Simple graphics functions

The simplest approach is to write a graphics function just for its side-effect of producing graphical output (i.e., using grid graphics functions as described in Chapter 5). The first example will be a simple graphics function to produce an image. The code in Figure 7.2 provides code defining a function `grid.imageFun()` for this purpose.

This function takes arguments to describe the number of rows and columns in the image (`nrow` and `ncol`), the colors to use for each cell in the image (`cols`), and the order in which those colors should be applied to the cells (`byrow`). Output is produced by a call to the `grid.rect()` function (line 12), which draws a rectangle for each cell in the image.

This function can be used to draw an array of rectangles just like any other plotting function. An example usage is given in the following code. First, a set of grey-scale colors are defined (these will be used throughout the chapter).

```
> greys <- grey(0.5 + (rep(1:4, 4) - rep(0:3, each=4))/10)
```

```
 1 grid.imageFun <- function(nrow, ncol, cols,
 2                           byrow=TRUE) {
 3   x <- (1:ncol)/ncol
 4   y <- (1:nrow)/nrow
 5   if (byrow) {
 6     right <- rep(x, nrow)
 7     top <- rep(y, each=ncol)
 8   } else {
 9     right <- rep(x, each=nrow)
10     top <- rep(y, ncol)
11   }
12   grid.rect(x=right, y=top,
13     width=1/ncol, height=1/nrow,
14     just=c("right", "top"),
15     gp=gpar(col=NA, fill=cols),
16     name="image")
17 }
```

Figure 7.2

A grid.imageFun() function. This function draws an array of nrow by ncol rectangles filled with the specified colors.

Now two images are drawn with the same colors, but different byrow settings (see Figures 7.3a and 7.3b).

```
> grid.imageFun(4, 4, greys)
```

```
> grid.imageFun(4, 4, greys, byrow=FALSE)
```

There is an obvious deficiency in this function because it does not perform any checking of its arguments to ensure that the correct information is being passed to it. For example, there is no check that nrow and ncol are numeric values of length 1. In general, in order to reduce the size and complexity of the code chunks, the examples will leave out input-checking code. This issue is addressed more seriously in the context of developing new graphical objects in Section 7.3.3.

The grid.imageFun() example shows that it is quite straightforward to create a new graphics function that just produces output. However, there are three important things to keep in mind when writing such a function: other people might want to embed your function as an element within a more complex scene; other people might want to embed more output within the output from your function; and other people might want to interactively modify the output

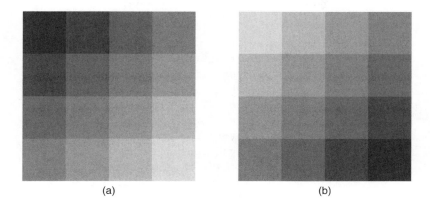

(a)   (b)

Figure 7.3
Output from the `grid.imageFun()` function. The two images use the same set of colors, but have different orientations. Image (a) has `byrow=TRUE` and image (b) has `byrow=FALSE`.

from your function. The following sections look at how you should design your function so that these tasks are straightforward for other people.

7.2.1   Embedding graphical output

The grid system is designed to allow graphical output to be embedded within other graphical output. All drawing occurs within the current viewport and no assumptions are made about the position or size of that viewport. New grid functions should be written with this in mind and it should not be assumed that output is being drawn into the entire device.

The `grid.imageFun()` function demonstrates this idea; this function just draws rectangles within the current viewport, wherever that may be and however large it may be.

On the other hand, it is sometimes important to enforce certain constraints on how graphical output is drawn. A good example is in the drawing of maps. Usually, a map is drawn with a specific aspect ratio so that, for example, 1 unit in the x-dimension has the same physical size as 1 unit in the y-dimension. In such cases, it may be necessary for a function to push its own viewports to enforce an aspect ratio before performing any drawing. A function to draw a map of Australia will be developed in order to demonstrate this idea.

The package `oz`[63] provides data for drawing maps of Australia. The

```
 1 grid.ozFun <- function(ozRegion) {
 2   pushViewport(
 3     viewport(name="ozlay",
 4              layout=grid.layout(1,1,
 5                                 widths=diff(ozRegion$rangex),
 6                                 heights=diff(ozRegion$rangey),
 7                                 respect=TRUE)))
 8   pushViewport(viewport(name="ozvp",
 9                         layout.pos.row=1,
10                         layout.pos.col=1,
11                         xscale=ozRegion$rangex,
12                         yscale=ozRegion$rangey,
13                         clip=TRUE))
14   index <- 1
15   for(i in ozRegion$lines) {
16     grid.lines(i$x, i$y, default.units="native",
17               name=paste("ozlines", index, sep=""))
18     index <- index + 1
19   }
20   upViewport(2)
21 }
```

Figure 7.4
A grid.ozFun() function. This function draws a map of Australia or some part
thereof.

ozRegion() function in the oz package returns an object of class "ozRegion"
containing x-axis and y-axis ranges, and a list of x-locations and y-locations
to draw map lines. The grid.ozFun() shown in Figure 7.4 makes use of
ozRegion() to draw a map of Australia using grid.

The most important part of this function is the pushing of viewports that
establish the correct aspect ratio for drawing the map (lines 2 to 13). The
first viewport contains a layout with a single cell set to the correct aspect ratio
and the second viewport occupies that cell and sets the appropriate "native"
coordinate system for the map. This allows the map to be drawn within any
viewport, but retain the appropriate shape.

The rest of the grid.ozFun() function just draws the lines representing the
Australian coastline (and state boundaries) using grid.lines().

The following code shows an example of the grid.ozFun() function being
used to draw all of Australia (see Figure 7.5). The map is not distorted even
though the region it is drawn in (indicated by the grey rectangle) is very wide.

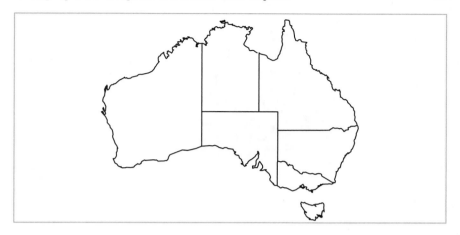

Figure 7.5

Example output from `grid.ozFun()`. By default it draws all of Australia.

```
> grid.ozFun(ozRegion())
```

7.2.2   Facilitating annotation

In addition to being able to produce graphical output within any context, it is vital that further graphical output can be added to the output of a graphical function. Again, the grid system is designed to facilitate this, by allowing navigation between viewports.

In this context, there are two important features of the `grid.ozFun()` function defined in Figure 7.4: the viewports that are pushed have names, `"ozlay"` and `"ozvp"` (lines 3 and 8); and the function calls `upViewport()` (not `popViewport()`) when it has finished drawing (line 20). These features mean that the viewports are available and accessible for other code to use after the `grid.ozFun()` function has done its drawing.

The following code provides an example of annotation using the `grid.imageFun()` function to add an image to output from the `grid.ozFun()` function (see Figure 7.6). In this example, only a small part of the South Australian coastline is used (the coastline close to the area where fluorescence data were gathered).

First of all, the latitude and longitude ranges are set up for the map (`mapLong` and `mapLat`) and for the image (`imageLong` and `imageLat`). Also, the set of colors for the image are calculated (`imageCols`). The prediction surface to be

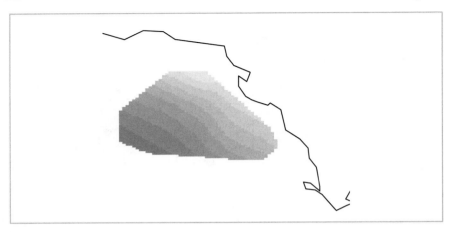

Figure 7.6
Annotating `grid.ozFun()` output. An image has been added using the
`grid.imageFun()` function.

plotted is in a variable called `fluoro.predict`, which has components x, y,
and z for the longitude, latitude, and predicted fluorescence value respectively.
These ranges and colors will be used throughout the rest of the chapter.

```
> mapLong <- c(132, 136)
> mapLat <- c(-35, -31.5)
> imageLong <- range(fluoro.predict$x)
> imageLat <- range(fluoro.predict$y)
> zbreaks <- seq(min(fluoro.predict$z, na.rm=TRUE),
                 max(fluoro.predict$z, na.rm=TRUE),
                 length=10)
> zcol <- cut(fluoro.predict$z, zbreaks,
              include.lowest=TRUE, labels=FALSE)
> ozgreys <- grey(0.5 + 1:9/20)
> imageCols <- ozgreys[zcol]
```

Now, the map and image can be drawn. The map is drawn first which produces
the coast line of South Australia and sets up the viewports `"ozlay"` and
`"ozvp"`.

```
> grid.ozFun(ozRegion(xlim=mapLong, ylim=mapLat))
```

The function `downViewport()` is used to navigate down to the viewport
`"ozvp"`, which has scales set up representing the latitude and longitude of

the map. This is only possible because the `grid.ozFun()` function specified useful names for the viewports it set up.

```
> downViewport("ozvp")
```

A further viewport is pushed to occupy the region where the image should be drawn and the image is drawn within that viewport.

```
> pushViewport(viewport(y=min(imageLat),
                        height=abs(diff(imageLat)),
                        x=max(imageLong),
                        width=abs(diff(imageLong)),
                        default.units="native",
                        just=c("right", "bottom")))
> grid.imageFun(50, 50, col=imageCols)
> upViewport(0)
```

7.2.3 Editing output

In addition to being able to add further output to a plot, it is useful to make it easy for others to modify the existing elements of a plot. The important step in this case is to provide a name for each piece of graphical output that your function produces.

The `grid.imageFun()` function uses the name `"image"` for the set of rectangles that it draws (line 16 in Figure 7.2) and the `grid.ozFun()` function names each map border that it draws `"ozlines`i`"`, where i varies from 1 to the number of borders drawn (line 17 in Figure 7.4).

These names are useful for interacting with the output from these functions, particularly for the purpose of editing the output. The following code presents a couple of examples of modifying the plot produced in Figure 7.6. The first edit reverses the set of colors used in the image. The second edit changes the color of all map borders to grey and makes the borders thicker (see Figure 7.7).

```
> grid.edit("image", gp=gpar(fill=rev(ozgreys)[zcol]))
> grid.edit("^ozlines[0-9]+$", gp=gpar(col="grey", lwd=2),
            grep=TRUE, global=TRUE)
```

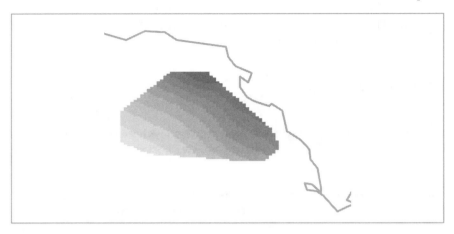

Figure 7.7
Editing `grid.ozFun()` output. Compared to Figure 7.6, the colors of the image have
been reversed and the Australian coastline is thicker and colored grey.

7.2.4 Absolute versus relative sizes

Another thing to consider when designing a graphics function is whether to
use absolute or relative coordinate systems and graphical parameters for sizing
graphical output. If absolute coordinates systems such as `"inches"`, `"cm"`, or
`"mm"` are used to size output, then the output will remain that size no matter
how large or small the surrounding viewport is made. This is also true of
graphical parameters such as `fontsize` (which specifies the size of text in
points), and `lwd`. If, on the other hand, relative coordinates such as `"npc"` or
`"native"` are used for sizing output, the output will resize with its container.
Graphical parameters that are relative like this are `cex` for sizing text and
`lex` for line width.

In general, absolute sizes are more appropriate for producing or fine-tuning a
piece of output for a specific use (e.g., a figure in an article). Relative sizes
are more appropriate when designing general graphics functions for others to
use, where it is unknown how large the final output will be. One possible
exception to this rule is the sizing of text. It is reasonable to set text size in
absolute terms (i.e., a particular point size) in order to ensure that the text
is legible.

The coordinate systems used for `"char"`, `"lines"`, `"strwidth"`, or
`"grobwidth"` units depend on the size of other output and so are considered
to be relative.

7.3 Graphical objects

A properly written graphics function can be very useful if it can be reused in other plots and arbitrarily added to or modified as described in previous sections. There are, however, a number of benefits to be gained from also creating a graphical object, or grob, to represent the output that your function produces.

The following sections consider again the development of functions to produce maps and images, but this time with an emphasis on creating *objects* that represent the output, rather than just producing output.

Defining new grobs involves working with classes and generic functions. This section assumes a familiarity with the basic ideas of object-oriented programming and its implementation in S3 classes and methods (see Section A.4 for a very brief introduction).

The design of classes and methods is a reasonably sophisticated process, there are often a number of possible designs to choose from, and it can be difficult to determine a "best solution." This means that it is impossible to provide a single definitive statement about how a new graphical object should be developed. Instead, this section presents a number of examples with several different implementations and there is a discussion of the advantages and disadvantages of different approaches.

7.3.1 Overview of creating a new graphical class

There are two main steps involved in defining a new graphical class. First of all, the structure of the class must be described. This consists of specifying the components of the new class — the information that is stored in objects of that class. For example, the `"rect"` class has components x, y, `width`, `height`, and `just` that describe the location and size of the rectangle.

The functions `grob()` and `gTree()` are used to define the structure for a new graphical class (as described in more detail in the next section). These functions ensure that all grobs have a number of standard components. For example, all grobs must have `gp`, `vp`, and `name` components. In addition, all classes derived from `"gTree"` (via the `gTree()` function) also have components `children` and `childrenvp` that describe the children of the gTree and how those children are drawn (see Section 7.3.4).

The second step in defining a new graphical class is to define the behavior of the class. This consists of writing methods for several important generic functions. Methods can be written to control the validation of a grob, how a grob is drawn, and what happens when a grob is modified. It is also possible to write methods for calculating the size of a grob. These generic functions are described in Sections 7.3.4 to 7.3.7.

7.3.2   Defining a new graphical class

The code in Figure 7.8 gives an example of defining a new graphical class. An `"imageGrob"` class is defined, which contains a description of the image output generated by the `grid.imageFun()` function that was defined earlier.

The `imageGrob()` function calls the function `gTree()` to create an object of a new class, `"imageGrob"`. An `imageGrob` is a gTree with several components that provide a high-level description of the an image (`ncol`, `nrow`, `cols`, and `byrow`). There is also a single child, which is a `rect` grob, representing the rectangles that will be drawn to produce the image. The `imageGrob()` function also provides the standard `gp`, `vp`, and `name` components, which should be available for all grobs.

The `makeImageRect()` function generates a `rect` grob from a high-level image description. This is very similar to the function `grid.imageFun()`, but it produces an object containing a description of some rectangles rather than drawing the rectangles and it calls `rectGrob()` rather than `grid.rect()` (line 11). This function is not intended to be used directly — it is just a "helper function" for the main `imageGrob()` function. This is an example of modular code that makes it easier to read the main function and it will be used later when some other class examples are considered.

The `grid.imageGrob()` function is just a convenience for producing graphical output from an `imageGrob` grob; it just creates an appropriate grob and draws it. The following code produces the same result as Figure 7.3a.

```
> grid.imageGrob(4, 4, greys)
```

There are now functions that define a new class and create an object of that class. Sections 7.3.3 to 7.3.7 describe how to define appropriate behavior for the new class so that it draws correctly and responds appropriately to being modified.

Summary of creating a new graphical class

A new class is derived from the `"grob"` class using the `grob()` function, or from the `"gTree"` class using the `gTree()` function (e.g., line 20 in Figure

```
1  makeImageRect <- function(nrow, ncol, cols, byrow) {
2    xx <- (1:ncol)/ncol
3    yy <- (1:nrow)/nrow
4    if (byrow) {
5      right <- rep(xx, nrow)
6      top <- rep(yy, each=ncol)
7    } else {
8      right <- rep(xx, each=nrow)
9      top <- rep(yy, ncol)
10   }
11   rectGrob(x=right, y=top,
12            width=1/ncol, height=1/nrow,
13            just=c("right", "top"),
14            gp=gpar(col=NULL, fill=cols),
15            name="image")
16 }

18 imageGrob <- function(nrow, ncol, cols, byrow=TRUE,
19                       name=NULL, gp=NULL, vp=NULL) {
20   igt <- gTree(nrow=nrow, ncol=ncol,
21                cols=cols, byrow=byrow,
22                children=gList(makeImageRect(nrow, ncol,
23                                             cols, byrow)),
24                gp=gp, name=name, vp=vp,
25                cl="imageGrob")
26   igt
27 }

29 grid.imageGrob <- function(...) {
30   igt <- imageGrob(...)
31   grid.draw(igt)
32 }
```

Figure 7.8

An "imageGrob" class. This is a grob-based equivalent of grid.imageFun().

7.8). This will ensure standard behavior when drawing and editing grobs, calculating the size of grobs, and so on (Sections 7.3.3 to 7.3.7 provide detailed information about the default behavior of grobs). Apart from the `cl` argument that specifies the name of the new class (line 25 in Figure 7.8), the arguments to these functions provide a list of components for the new class.

There are some standard components common to all grobs: `gp`, `vp`, and `name`. It is sensible to make these available via the constructor function for your new class (e.g., line 19 in Figure 7.8).

The `gp` component is designed to contain a `gpar` object, which is a set of graphical parameter settings; the `vp` component is designed to hold a viewport or a viewport path; and the `name` component provides the name for the grob. All of these are validated automatically and are used in the drawing and editing of the grob (see Section 7.3.3).

The `"gTree"` class defines two more standard components: the `children` component contains the children of the gTree (as a gList) and the `childrenvp` component contains viewports for the children to be drawn within (used in the drawing of the children of the gTree). An example of the use of the `children` component is shown in Figure 7.8 on line 22.

All other components are at the discretion of the class designer.

Having defined a new graphical class, it is then necessary to write one or more methods for some important generic functions as described in the following sections.

7.3.3   Validating grobs

This section describes the `validDetails()` function, which is important for ensuring that the components of a grob contain valid values.

The code examples used in the examples of simple graphics functions ignored the issue of checking user input to ensure that valid values are supplied for arguments or components. This issue becomes particularly important when dealing with grobs because it is not only possible to supply invalid values when a grob is first created, but also whenever a grob is modified via `grid.edit()` or `editGrob()`.

Default validating behavior

When a grob is created or modified, it is automatically validated. The validation checks that the `gp`, `vp`, and `name` components of a grob are sensible (and for a gTree, the `children` and `childrenvp` components are also checked) and

```
 1 validDetails.imageGrob <- function(x) {
 2   if (!is.numeric(x$nrow) || length(x$nrow) > 1 ||
 3       !is.numeric(x$ncol) || length(x$ncol) > 1)
 4     stop("nrow and ncol must be numeric and length 1")
 5   if (!is.logical(x$byrow))
 6     stop("byrow must be logical")
 7   x
 8 }

10 validDetails.ozGrob <- function(x) {
11   if (!inherits(x$ozRegion, "ozRegion"))
12     stop("Invalid ozRegion")
13   x
14 }
```

Figure 7.9
Some validDetails() methods. These are called when an imageGrob or an ozGrob
is first created, or when such an object is modified using grid.edit().

then the validDetails() generic function is called. By default this function
does nothing. A new class should define a method to check the components
that are specific to that class.

The imageGrob example

Figure 7.9 shows a validDetails() methods for the "imageGrob" class (there
is also a method for the "ozGrob" class, which is defined in the next section).
The validDetails() method for the "imageGrob" class (lines 1 to 8) checks
that the nrow and ncol components are numeric and of length 1 and that
the byrow component is a logical vector. The return value of the method is
the validated imageGrob (line 7). All validDetails() methods must do this
whether they modify the grob or not.

With these validation methods defined, both the creation and the modification of an imageGrob will perform checks to ensure that the components of
the imageGrob contain valid values. The following code demonstrates the validation at work. First of all, the creation of an imageGrob fails because byrow
is not a logical value.

```
> grid.imageGrob(4, 4, greys, byrow="what?")

Error in validDetails.imageGrob(x) :
    byrow must be logical
```

In this next example, an `imageGrob` is created with valid components, but then it is edited with an invalid `nrow` component specification.

```
> ig <- imageGrob(4, 4, greys)
```

```
> editGrob(ig, nrow="what?")
```

Error in validDetails.imageGrob(x) :
    nrow and ncol must be numeric and length 1

7.3.4   Drawing grobs

This section describes the `drawDetails()` generic function, which is important for ensuring that appropriate output is produced when a grob is drawn.

The function `grid.draw()` produces graphical output from a grob and there is default drawing behavior for all grobs.

Default drawing behavior

By default, if the `vp` component of a grob contains a viewport (or viewport stack, list, or tree), the viewport is pushed before any output is produced. If the `vp` component is a `vpPath`, the path is used to navigate down to the relevant viewport using `downViewport()`.* Also, if the `gp` components contains a `gpar` object, those graphical settings are enforced before any output is produced.

After the output has been produced, graphical settings are reverted and any viewports that were pushed are popped. If there was navigation down to a viewport, then that navigation is reversed with a call to `upViewport()`.

By default, no graphical output is produced for a grob. The generic function `drawDetails()` is called so that classes can define a method that calls grid functions to produce output. For classes derived from `"gTree"`, as well as calling the `drawDetails()` function, all of the children of the gTree are drawn (by calling `grid.draw()` for each child).

The default behavior for grobs also takes care of recording the grob on the grid display list.

---

*The navigation down is performed with `strict=TRUE`.

A further default behavior for gTrees is that, before drawing its children, a gTree will push any viewports in its `childrenvp` component, then navigate back up again. This means that all viewports in the `childrenvp` component are available for the grobs in the `children` component of the gTree to navigate down to. An example is given later (see the sub-section "An `ozGrob` example" and Figure 7.10).

The `imageGrob` example

The `grid.imageGrob()` function in Figure 7.8 draws an image by creating an `imageGrob` object and calling `grid.draw()`. This provides the same interface as the `grid.imageFun()` function and produces exactly the same output. The following code shows a simple example. The output is exactly the same as Figure 7.3a.

```
> grid.imageGrob(4, 4, greys)
```

The output is generated automatically in the `imageGrob` case because of the default drawing behavior defined for gTrees. When a gTree is drawn, it draws all of its children, so when an `imageGrob` is drawn, the `rect` grob child is drawn automatically.

The following code demonstrates the other automatic drawing behavior for grobs (pushing of the `vp` component and enforcing `gp` settings). The code draws an image like the previous one, except that the drawing occurs within a viewport specified via the `vp` component and the entire image is made transparent via the `gp` component.*

```
> grid.imageGrob(4, 4, greys, name="imageGrob",
               vp=viewport(width=0.5, height=0.5),
               gp=gpar(alpha=0.5))
```

An `ozGrob` example

The `"imageGrob"` class provides a simple example of the default drawing behavior for a gTree with children. The code in Figure 7.10 shows the definition of an `"ozGrob"` class, which will be used to demonstrate the drawing of a gTree with children *and* a `childrenvp` component.

---

*This example will not work on many graphics devices because they do not support transparency and will respond to a non-opaque color by not drawing anything! The PDF and the Quartz devices should behave sensibly.

```
 1 makeOzViewports <- function(ozRegion) {
 2   vpStack(viewport(name="ozlay", layout=grid.layout(1, 1,
 3                      widths=diff(ozRegion$rangex),
 4                      heights=diff(ozRegion$rangey),
 5                      respect=TRUE)),
 6           viewport(name="ozvp", layout.pos.row=1,
 7                      layout.pos.col=1,
 8                      xscale=ozRegion$rangex,
 9                      yscale=ozRegion$rangey,
10                      clip=TRUE))
11 }

13 makeOzLines <- function(ozRegion) {
14   numLines <- length(ozRegion$lines)
15   lines <- vector("list", numLines)
16   index <- 1
17   for(i in ozRegion$lines) {
18     lines[[index]] <- linesGrob(i$x, i$y,
19                        default.units="native",
20                        vp=vpPath("ozlay", "ozvp"),
21                        name=paste("ozlines", index, sep=""))
22     index <- index + 1
23   }
24   do.call("gList", lines)
25 }

27 ozGrob <- function(ozRegion, name=NULL, gp=NULL, vp=NULL) {
28   gTree(ozRegion=ozRegion, name=name, gp=gp, vp=vp,
29     childrenvp=makeOzViewports(ozRegion),
30     children=makeOzLines(ozRegion),
31     cl="ozGrob")
32 }

34 grid.ozGrob <- function(...) {
35   grid.draw(ozGrob(...))
36 }
```

Figure 7.10

An "ozGrob" class. This is a grob-based equivalent of grid.ozFun().

In this example, the `makeOzViewports()` function and the `makeOzLines()` function are both just helper functions. The functions `ozGrob()` and `grid.ozGrob()` are the only functions that other people will use.

An `ozGrob` is a gTree with a single component, `ozRegion`, containing a description of the region of Australia to map (line 28 in Figure 7.10). An `ozGrob` also has a number of children, all of which are `lines` grobs representing the coastline and state boundaries to draw (line 30), and an `ozGrob` has a viewport stack in its `childrenvp` component (line 29). These viewports create a region with the right aspect ratio for drawing a map and the children of the `ozGrob` are all created with viewport paths to specify that they should be drawn within this region (line 20).

When an `ozGrob` is drawn, the viewports in its `childrenvp` component are pushed as part of the default drawing behavior for gTrees, then the grobs in its `children` component are drawn. Each child has a `vp` component indicating which viewport to navigate to before drawing.

The `grid.ozGrob()` function is just a convenient front-end for drawing an `ozGrob`. This can be used just like the function `grid.ozFun()` to draw some or all of a map of Australia. The following code produces exactly the same output as shown in Figure 7.5. There are more examples using `ozGrob` objects in later sections.

```
> grid.ozGrob(ozRegion())
```

An `ozImage` example

Both the `"imageGrob"` class and the `"ozGrob"` class are derived from the `"gTree"` class. This means that they have other grobs as children and the default drawing behavior for gTrees draws those children correctly when the `imageGrob` or `ozGrob` is drawn. This section looks at an example where a `drawDetails()` method has to be written in order to produce any output.

A typical reason for needing to write a `drawDetails()` method is that your new class does not have a fixed set of grobs as children. Axes that must calculate tick marks on the fly are a good example (it is only possible to figure out how many tick marks to draw and where to locate them when the axis is actually drawn). Sometimes, it may be necessary because the children of a gTree need to be combined in a complex way.

In order to demonstrate the definition of a `drawDetails()` method, an `"ozImage"` class will be defined. This class combines an `ozGrob` and an `imageGrob` and has to do the drawing itself to get them combining correctly (for producing output like that in Figure 7.6).

```
1 ozImage <- function(mapLong, mapLat,
2                      imageLong, imageLat, cols) {
3   grob(mapLong=mapLong, mapLat=mapLat,
4        imageLong=imageLong, imageLat=imageLat, cols=cols,
5        cl="ozImage")
6 }

8 drawDetails.ozImage <- function(x, recording) {
9   grid.draw(ozGrob(ozRegion(xlim=x$mapLong,
10                             ylim=x$mapLat)))
11   depth <- downViewport(vpPath("ozlay", "ozvp"))
12   pushViewport(viewport(y=min(x$imageLat),
13                         height=diff(range(x$imageLat)),
14                         x=max(x$imageLong),
15                         width=diff(range(x$imageLong)),
16                         default="native",
17                         just=c("right", "bottom")))
18   grid.draw(imageGrob(50, 50, col=x$col))
19   popViewport()
20   upViewport(depth)
21 }
```

Figure 7.11
An "ozImage" class. This combines an imageGrob with an ozGrob to make a larger,
more complex grob.

The code in Figure 7.11 shows the definition of an "ozImage" class. An
ozImage is just a grob (it has no children; lines 3 to 5) so without a
drawDetails() method, it would produce no output. In order to produce
output when an ozImage is drawn, a drawDetails() method is defined for
the "ozImage" class (lines 8 to 21). This method creates an ozGrob and
draws it (lines 9 to 10), then navigates down to the "ozvp" viewport, pushes
a viewport within which to draw the image, creates and draws an imageGrob
(line 18), and finally navigates back up to the viewport that it started in.

With this class defined, Figure 7.6 can be produced as follows.

```
> grid.draw(ozImage(mapLong, mapLat,
                     imageLong, imageLat, imageCols))
```

An important point about drawDetails() methods is that none of the drawing and viewport operations within a drawDetails() method are recorded

on the display list.* For example, as a result of the above code, there is an `ozImage` grob on the display list, but there is neither an `ozGrob` nor an `imageGrob` on the display list. This has implications for editing output which are discussed in' the next section.

7.3.5   Editing grobs

This section describes the `editDetails()` generic function, which is important for ensuring that a grob responds appropriately when it is edited. It is particularly important for classes derived from `"gTree"` to ensure that the children of the gTree are updated when the high-level components of the gTree are modified.

One advantage of defining a grob to represent graphical output is that the grob provides a high-level interface to the graphical output. For example, an `imageGrob` contains components that describe an image in terms of how many rows and columns it has. The low-level description of the precise location of individual rectangles within the image is left to the lower-level `rect` grob. This means that it is possible to modify the high-level description in order to change the graphical output. For example, the number of rows in an `imageGrob` could be modified simply by changing the high-level `nrow` component rather than by having to modify the location and size of all of the low-level rectangles. Unfortunately, modifying the high-level description of an `imageGrob` as it has been defined so far will have no effect on the output because it will have no effect on the children of the `imageGrob`. For gTrees with children, it is necessary to provide instructions for how to change the children when the high-level description is modified.

Default editing behavior

The `grid.edit()` function and the `editGrob()` function are used to modify a grob.

When a grob is modified, the components of the grob are set to the new values and the `editDetails()` generic function is called. The default behavior is to do nothing, but a class can define a method which, for example, propagates a change in its components to its children.

---

*Prior to R version 2.1.0, this was not the case and recording had to be explicitly turned off by specifying `recording=FALSE` in all drawing and viewport operations.

The `imageGrob` example

In the case of an `imageGrob`, an `editDetails()` method is required to en-sure that the child `rect` corresponds to the high-level description in the `imageGrob`. Figure 7.12 shows code defining an `editDetails()` method for the `"imageGrob"` class (lines 1 to 10). This method totally recreates the child `rect` grob if any of the `ncol`, `nrow`, or `byrow` arguments are modified (lines 2 to 5), and edits the child `rect` grob if the `cols` argument is modified (lines 6 to 8). A very important feature of the function is that it returns the modified grob (line 9). All `editDetails()` methods must do this.

The `ozGrob` example

Figure 7.12 also shows an `editDetails()` method for the `"ozGrob"` class. This method ensures that changes to the `ozRegion` component of an `ozGrob` will be reflected in the children of the `ozGrob` by completely recreating the `childrenvp` and `children` components of the `ozGrob`.

The `ozImage` example

There is no method for the `"ozImage"` class because it is only a grob and there are no children to propagate changes to. The output of an `ozImage` is recreated by its `drawDetails()` method whenever an `ozImage` is edited.

The `imageGrob` example again

With the `editDetails()` method defined for the `"imageGrob"` class, it is possible to edit an `imageGrob`. The following code creates an image (see Figure 7.13a) and then modifies the orientation via the high-level descrip-tion in the `imageGrob`. The changes are passed on to the `rect` child by the `editDetails()` method (see Figure 7.13b).

```
> grid.imageGrob(4, 4, greys, name="imageGrob")

> grid.edit("imageGrob", byrow=FALSE)
```

A gTree with child grobs not only allows interaction with a high-level descrip-tion of graphical output, but it also makes it possible to access the low-level description as well through the child grobs. For example, it is possible to edit the low-level `rect` child of an `imageGrob`. The following code modifies the image drawn in the previous example to change the borders of the child rectangles to be white and very wide (see Figure 7.13c). The only important

```
 1 editDetails.imageGrob <- function(x, specs) {
 2   if (any(c("ncol", "nrow", "byrow") %in% names(specs))) {
 3     x <- addGrob(x, makeImageRect(x$nrow, x$ncol,
 4                                   x$cols, x$byrow))
 5   }
 6   if (any(c("cols") %in% names(specs))) {
 7     x <- editGrob(x, "image", gp=gpar(fill=x$cols))
 8   }
 9   x
10 }

12 editDetails.ozGrob <- function(x, specs) {
13   if ("ozRegion" %in% names(specs)) {
14     x$childrenvp <- makeOzViewports(x$ozRegion)
15     x <- setChildren(x, makeOzLines(x$ozRegion))
16   }
17   x
18 }
```

Figure 7.12
Some editDetails() methods for imageGrob and ozGrob objects. These will be run
when such objects are modified using grid.edit().

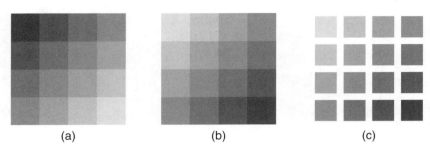

(a)   (b)   (c)

Figure 7.13

Editing an imageGrob. Panel (a) shows a simple imageGrob. In panel (b), the imageGrob has been modified by changing the byrow argument. In panel (c), the rect grob within the imageGrob has been modified to change the borders of each rectangle in the image to be thick and white.

design aspect here is that the "imageGrob" class provides a name, "image", for the rect child. This makes it possible to specify a gPath to the rect.

```
> grid.edit("imageGrob::image", gp=gpar(col="white", lwd=6))
```

The difference between editing the low-level rect object and recreating it is that when the rect object is edited it will retain any other customizations that have been made to it. The following code gives a small example. First of all an imageGrob object is drawn (the output is identical to Figure 7.13a).

```
> grid.imageGrob(4, 4, greys, name="imageGrob")
```

In the first edit, the low-level rect object is edited so that the borders of the individual rectangles are drawn white (see Figure 7.14a).

```
> grid.edit("imageGrob::image", gp=gpar(col="white"))
```

The second edit modifies the high-level cols component in the imageGrob object, but because this only edits the low-level rect object the individual rectangle borders are retained (see Figure 7.14b).

```
> grid.edit("imageGrob", cols=rev(greys))
```

In the final edit, the high-level byrow argument is edited, which causes the low-level rect object to be recreated and the individual borders are lost (see Figure 7.14c).

```
> grid.edit("imageGrob", byrow=FALSE)
```

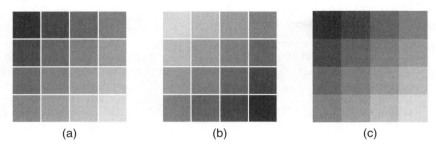

(a)   (b)   (c)

Figure 7.14
Low-level editing of an `imageGrob`. Panel (a) shows an `imageGrob` with the `rect` grob modified so that all of the rectangles in the image have a white border. In panel (b), the `cols` argument of the `imageGrob` has been modified, which changes the fill color of each rectangle, but does not alter the borders (which are still white). In panel (c), the `byrow` argument of the `imageGrob` has been modified and this causes the `rect` grob to be remade so the white borders are lost.

The `ozImage` example again

It is worth noting that for grobs that do not have any children, but produce output by creating and drawing grobs in a `drawDetails()` method, it is not actually possible to perform this sort of low-level editing. For example, it is not possible to edit the low-level `imageGrob` or `ozGrob` of an `ozImage` because the `imageGrob` and `ozGrob` are never stored anywhere. Some solutions to this problem are discussed in Section 7.3.10.

7.3.6   Sizing grobs

This section describes the `widthDetails()` and `heightDetails()` generic functions, which are useful for calculating the amount of space that a grob requires for drawing.

Every new graphical class should have appropriate behavior defined for validating, drawing, and editing objects of the class. This section looks at defining behavior for querying a grob for its width or height, which only makes sense in some cases.

This is most often used for providing column widths and row heights for grid layouts, but can also be used to position one grob relative to another (e.g., to draw a rectangular border around a piece of text).

Default sizing behavior

The calculation of "grobwidth" and "grobheight" units include calls to the generic functions, widthDetails() and heightDetails() respectively, which should return a unit describing the width and height of a grob.

The default widthDetails() and heightDetails() methods return unit(1, "null"). Within layouts this means that a grob gets an equal share of the available space (see Section 5.5.6). Outside of a layout, this converts to a size of zero. There are predefined widthDetails() and heightDetails() methods for most primitives and for the "frame" class (see Section 6.4).

A ribbonLegend example

In order to demonstrate the use of these methods, a "ribbonLegend" class will be defined (see Figures 7.15 and 7.16).

A ribbonLegend consists of several components describing the number of levels to represent in the legend, the colors to use for each level (line 6 in Figure 7.16). It is also possible to control the amount of empty space to leave around the legend (line 3). The children of a ribbonLegend are a rect grob to draw rectangles for the levels and a lines grob and a text grob to show the legend scale (see the ribbonKids() function in Figure 7.15). There is a childrenvp component that contains a vpTree: the parent viewport defines a layout and then two child viewports occupy column 2 and column 3 of row 2 of that layout (see the ribbonVps() function in Figure 7.15). The layout is created with column widths based on the space needed for the scale labels (in column 3), one line of text for the "ribbon" of rectangles (in column 2), and the required empty space around the outside (columns 1 and 4).

A widthDetails() method is defined for the "ribbonLegend" class in Figure 7.16 (lines 12 to 14). The width of a ribbonLegend is calculated as the sum of the widths of the layout in the top viewport of the childrenvp component (line 13), which reflects the space required to draw the legend.

This class can be used to produce a ribbon legend as shown at the right-hand side of Figure 7.1. If a ribbon legend is used as the data for a "grobwidth" unit, it will request enough space to draw itself. Examples are given in Sections 7.3.8 and 7.3.9.

7.3.7   Pre-drawing and post-drawing

There are two more generic functions that have not yet been mentioned: preDrawDetails() and postDrawDetails(). These functions are called as

```
1 calcBreaks <- function(nlevels, breaks, scale) {
2   if (is.null(breaks)) {
3     seq(min(scale), max(scale), diff(scale)/nlevels)
4   } else {
5     breaks
6   }
7 }

9 ribbonVps <- function(nlevels, breaks, margin, scale) {
10   breaks <- format(signif(calcBreaks(nlevels, breaks, scale),
11                           3))
12   vpTree(
13     viewport(name="layout", layout=
14       grid.layout(3, 4,
15         widths=unit.c(margin, unit(1, "lines"),
16                       max(unit(0.8, "lines") +
17                       stringWidth(breaks)), margin),
18         heights=unit.c(margin, unit(1, "null"), margin))),
19     vpList(viewport(layout.pos.col=2, layout.pos.row=2,
20                     yscale=scale, name="ribbon"),
21            viewport(layout.pos.col=3, layout.pos.row=2,
22                     yscale=scale, name="labels")))
23 }

25 ribbonKids <- function(nlevels, breaks, cols, scale) {
26   breaks <- calcBreaks(nlevels, breaks, scale)
27   nb <- length(breaks)
28   tickloc <- breaks[-c(1, nb)]
29   gList(rectGrob(y=unit(breaks[-1], "native"),
30                  height=unit(diff(breaks), "native"),
31                  just="top", gp=gpar(fill=cols),
32                  vp=vpPath("layout", "ribbon")),
33         segmentsGrob(x1=unit(0.5, "lines"),
34                      y0=unit(tickloc, "native"),
35                      y1=unit(tickloc, "native"),
36                      vp=vpPath("layout", "labels")),
37         textGrob(x=unit(0.8, "lines"),
38                  y=unit(tickloc, "native"),
39                  just="left",
40                  label=format(signif(tickloc, 3)),
41                  vp=vpPath("layout", "labels")))
42 }
```

Figure 7.15
Helper functions for a `"ribbonLegend"` class. The class itself is defined in Figure 7.16.

```
 1 ribbonLegend <- function(nlevels=NULL, breaks=NULL, cols,
 2                           scale=range(breaks),
 3                           margin=unit(0.5, "lines"),
 4                           gp=NULL, vp=NULL, name=NULL) {
 5   gTree(
 6     nlevels=nlevels, breaks=breaks, cols=cols, scale=scale,
 7     children=ribbonKids(nlevels, breaks, cols, scale),
 8     childrenvp=ribbonVps(nlevels, breaks, margin, scale),
 9     gp=gp, vp=vp, name=name, cl="ribbonLegend")
10 }

12 widthDetails.ribbonLegend <- function(x) {
13   sum(layout.widths(viewport.layout(x$childrenvp[[1]])))
14 }
```

Figure 7.16
A "ribbonLegend" class. This consists of a "ribbon" of rectangles filled with the
specified colors, plus an axis showing the scale.

part of the default drawing behavior (see Section 7.3.4).

Default pre/post-drawing behavior

The pushing and popping of viewports in vp components described
in the default drawing behavior (Section 7.3.4) includes a call to the
preDrawDetails() generic function (after the vp component has been
pushed, but before the drawDetails() method is called) and a call to the
postDrawDetails() generic function (after the drawDetails() method has
been called, but before the vp component is popped).

By default, the generic functions do nothing, but a new graphical class can
define a method to perform additional pushing, popping, and navigation of
viewports if required.

The pre-drawing and post-drawing are separate actions from the actual drawing because they are also performed during the evaluation of "grobwidth"
and "grobheight" units. This is done so that the size of a grob is calculated
based on the context in which it is drawn (i.e., so that the size corresponds
to the actual size of the graphical output). This means that any pushing
(and popping) of viewports for a new grob class must be performed in a
preDrawDetails() (and postDrawDetails()) method if they will have any
effect on the calculations on the size of the grob.

Summary of graphical object methods

Defining the behavior for a new graphical class requires writing one or more methods for the standard grid generic functions:

- **always** write a constructor function for the class to generate a grob or a gTree containing the description of what to draw.
- **always** write a `validDetails()` method for checking the validity of values in the non-standard components of the class.
- **sometimes** write a `drawDetails()` method to specify how to draw the class.
- **rarely** write `preDrawDetails()` and `postDrawDetails()` methods if the drawing involves pushing viewports that affect the determination of the size of the graphical output.
- **always** (**for gTrees**) write an `editDetails()` method so that changes to the high-level description are propagated to child grobs.
- **sometimes** write `widthDetails()` and `heightDetails()` methods if the size of the graphical output can be sensibly determined.

7.3.8 Completing the example

Almost all of the components required to produce Figure 7.1 have now been defined. The following code produces an `ozImage` that contains a portion of the South Australian coastline and an image representing fluorescence just off the coast.

```
> ozimage <- ozImage(mapLong, mapLat,
                     imageLong, imageLat, imageCols)
```

This next piece of code creates a `ribbonLegend` for the image plot.

```
> ribbonlegend <- ribbonLegend(breaks=zbreaks,
                               cols=ozgreys,
                               scale=range(zbreaks),
                               gp=gpar(cex=0.7))
```

The final piece required is the ability to draw a "key" consisting of a map of Australia with a rectangle to indicate the region drawn in the main image. The code in Figure 7.17 defines an `"ozKey"` class for this purpose. An `ozKey` is a gTree with an `ozGrob` and a `rect` grob as children. An `ozKey` draws its children within a viewport, the location and size of which are specified

```
1 ozKey <- function(x, y, width, height, just,
2                    mapLong, mapLat) {
3   gTree(childrenvp=viewport(name="ozkeyframe",
4                             x=x, y=y, just=just,
5                             width=width, height=height),
6         children=gList(ozGrob(ozRegion(), vp="ozkeyframe",
7                               gp=gpar(lwd=0.1)),
8                        rectGrob(x=mean(mapLong),
9                                 y=mean(mapLat),
10                                width=abs(diff(mapLong)),
11                                height=abs(diff(mapLat)),
12                                default.units="native",
13                                gp=gpar(lwd=1),
14                                vp=vpPath("ozkeyframe",
15                                          "ozlay", "ozvp"))))
16 }
```

Figure 7.17
An "ozKey" class. This consists of a map of Australia with a rectangle superimposed.

when the ozKey is constructed. A drawDetails() method is not required for this class because the default drawing behavior for gTrees is sufficient. A validDetails() method should be written and an editDetails() method would be required for an ozKey to respond properly to editing; however, these are left as an exercise for the reader.

The following code constructs an ozKey grob for the image plot.

```
> ozkey <- ozKey(x=unit(1, "npc") - unit(1, "mm"),
                 y=unit(1, "npc") - unit(1, "mm"),
                 width=unit(3.5, "cm"),
                 height=unit(2, "cm"),
                 just=c("right", "top"),
                 mapLong, mapLat)
```

Finally, the ozImage, ozKey, and ribbonLegend grobs are used to construct Figure 7.1 as follows.

```
> fg <- frameGrob()
> fg <- packGrob(fg, ozimage)
> fg <- placeGrob(fg, ozkey)
> fg <- packGrob(fg, ribbonlegend, "right")
> grid.draw(fg)
```

This makes use of the fact that the `ribbonLegend` will be allocated the correct width because it has a `widthDetails()` method defined.

7.3.9 Reusing graphical elements

Having created all of these graphical objects, there exists a set of graphical elements that produce useful graphical output, that can have other output added to them, that can be modified at different levels, *and* that can be embedded in other settings. This means that it is possible to use them to produce a quite different sort of plot. Figure 7.18 shows an example that combines an `ozGrob` and several `ribbonLegends` in a quite different way.

First of all, a map of Australia is drawn, then a `ribbonLegend` grob is drawn for each city to show the range of average monthly temperatures. The data are minimum and maximum monthly temperatures at six major cities spread around Australia available as the data set `ozTemp` in the `RGraphics` package.

```
> grid.ozGrob(ozRegion())
> downViewport("ozvp")
> for (i in 1:(dim(ozTemp)[1])) {
    grid.points(ozTemp$long[i], ozTemp$lat[i], pch=16)
    rl <- ribbonLegend(breaks=c(min(ozTemp$min),
                               ozTemp$min[i],
                               ozTemp$max[i],
                               max(ozTemp$max)),
                    cols=c("white", "grey", "white"))
    pushViewport(viewport(x=unit(ozTemp$long[i], "native"),
                          y=unit(ozTemp$lat[i], "native"),
                          height=unit(1, "inches"),
                          width=grobWidth(rl),
                          clip="off"))
    grid.circle(r=0.7,
                gp=gpar(col="grey", fill="white", alpha=0.8))
    grid.draw(rl)
    popViewport()
  }
> upViewport(0)
```

This example uses semitransparent colors so it will only reproduce properly on a PDF or Quartz device.

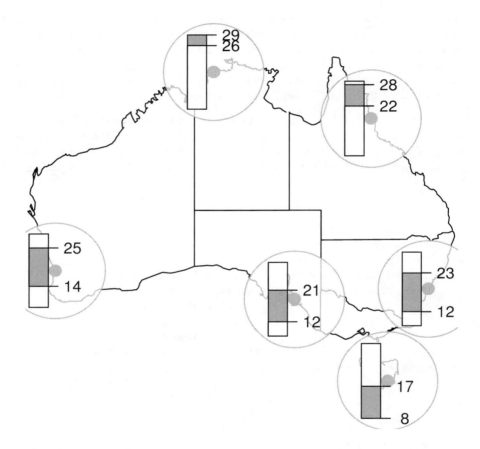

Figure 7.18
A plot of temperature data for several cities in Australia. This plot was composed
using the same `ozGrob` and `ribbonLegend` grobs that were used to construct the
oceanographic plot in Figure 7.1.

7.3.10 Other details

This section describes some more detailed and abstract issues that can arise in the design of a new graphical object.

Extending other grobs

From a design point of view, the correct implementation of the `"imageGrob"` class (Figure 7.8) probably would have been to derive it from the predefined `"rect"` class (rather than deriving it from the `"gTree"` class and having it *contain* a `rect` object as its child). In other words, from a design perspective, an `image` *is* a `rect`, simply with a different parameterization.

Unfortunately, this sort of implementation is awkward because grid uses the S3 class system, which does not support inheritance of structure. Deriving an `"image"` class from the `"rect"` grob class would mean that generic functions are automatically inherited, but the inheritance of structure (components) has to be done by hand. This requires knowledge of the internal structure of the `"rect"` class and will fail if there are any changes to that structure, so this approach is not recommended.

This means that the standard approach has to be to derive new classes directly from the `"grob"` class or the `"gTree"` class, not from other graphical classes.

Display lists

As mentioned in Section 1.3.3, R's graphics engine maintains a display list — a record of all graphical output on a device — and this is used to redraw a scene if a device is resized (among other things). The output from both traditional and grid graphics functions is recorded on this display list.

Grid also maintains its own separate display list, which is used for accessing grobs in the current scene and for redrawing the current scene after editing (i.e., after a call to `grid.edit()`). The grid display list can be replayed explicitly using the `grid.refresh()` function.

The grid display list can be disabled using `grid.display.list()`, which saves on grid's memory usage, but disables grid's ability to modify and redraw a scene. If the grid display list is disabled, the functions `grid.edit()`, `grid.get()`, `grid.add()`, and `grid.remove()` will no longer work.

It is possible to record grid output only on the grid display list with the `engine.display.list()` function, as shown by the following code. Redrawing will be slightly slower, but this avoids the memory cost of having output recorded on both the grid display list and the graphics engine display list.

```
> engine.display.list(FALSE)
```

This action does not turn off the graphics engine display list (so traditional graphics output is still recorded on the engine display list — see Section 1.3.3 for use of the `dev.control()` function to turn off the engine display list).

Calculations during drawing

Grid units and layouts make it possible to specify quite complex arrangements of output in a "declarative" manner. For example, the idea that a particular region should be square (have an aspect ratio of 1) can be expressed at a high level (e.g., by specifying both width and height as `unit(1, "snpc")`) and the system will ensure that this occurs. There is no need to calculate the physical dimensions of the current viewport and from those determine how to make a square region.

It is, however, sometimes necessary to perform calculations by hand. For example, consider the problem of splitting text into several lines based on the width of the available space. The code in Figure 7.19 defines a function, `splitString()`, to perform this operation (in a very simple-minded way). The important part of this function is the use of the `convertWidth()` function to obtain the size of the current line of text in inches (line 11) for comparison with the size of the current region in inches (lines 6 to 8).

The following code uses the `splitString()` function to draw some text within the current viewport (see the left-hand panel in Figure 7.20).

```
> text <- "The quick brown fox jumps over the lazy dog."
> grid.text(splitString(text),
            x=0, y=1, just=c("left", "top"))
```

There is a problem with the above code. If it is used to draw into a window and then the window is resized, the calculations are not rerun and the line splitting becomes incorrect. A similar problem occurs if the code is run on one device and then copied to another with different physical dimensions.

The issue is that only drawing actions are recorded on the display list, not any calculations leading up to the drawing. Anything that works off the display list (like redrawing after a resize) only reruns drawing actions.

There are two solutions to this problem. One solution rests on the fact that all code within a `drawDetails()` method (or a `preDrawDetails()` or `postDrawDetails()` method) is captured on the graphics engine display list.*

---

*Prior to R version 2.1.0, this was not the case. In earlier versions, in order to get

```
1 splitString <- function(text) {
2   strings <- strsplit(text, " ")[[1]]
3   newstring <- strings[1]
4   linewidth <- stringWidth(newstring)
5   gapwidth <- stringWidth(" ")
6   availwidth <-
7     convertWidth(unit(1, "npc"),
8                  "inches", valueOnly=TRUE)
9   for (i in 2:length(strings)) {
10    width <- stringWidth(strings[i])
11    if (convertWidth(linewidth + gapwidth + width,
12                     "inches", valueOnly=TRUE) <
13        availwidth) {
14      sep <- " "
15      linewidth <- linewidth + gapwidth + width
16    } else {
17      sep <- "\n"
18      linewidth <- width
19    }
20    newstring <- paste(newstring, strings[i], sep=sep)
21  }
22  newstring
23 }
```

Figure 7.19

A `splitString()` function. This function takes a piece of text and splits it into multiple lines so that the text will fit (horizontally) within the current viewport. Validation checks (e.g., whether **strings** is a character vector of length at least 2) have not been included.

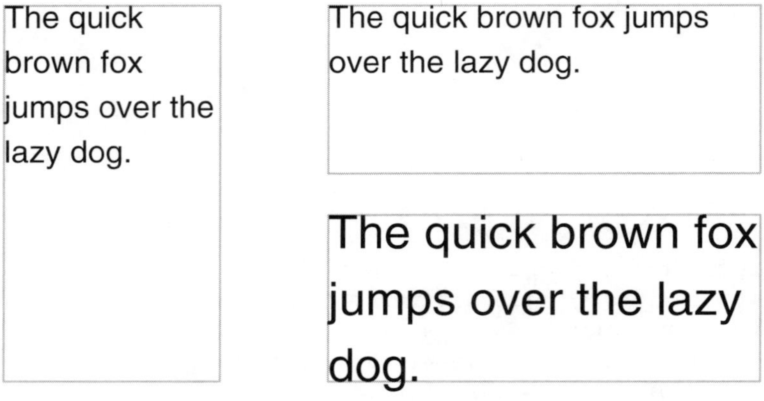

The quick brown fox jumps over the lazy dog.

The quick brown fox jumps over the lazy dog.

The quick brown fox jumps over the lazy dog.

Figure 7.20

Performing calculations before drawing. If the drawing of a grob depends on calculations (in this case, calculations to split text into multiple lines to fit horizontally within the current viewport), the calculations should be included within a
`drawDetails()` method. This means that the calculations will be rerun if the device
is resized (left panel versus top-right panel) or if the grob is edited to make the font
size larger (top-right panel versus bottom-right panel).

The code in Figure 7.21 uses this fact to create a `"splitText"` class with a
`drawDetails()` method that performs the calculations.*

A `splitText` grob will recalculate the line breaks when a device is resized or
on copies between devices (see the top-right panel of Figure 7.20).

```
> splitText <- splitTextGrob(text, name="splitText")
> grid.draw(splitText)
```

Another advantage of creating a grob with a `drawDetails()` method is that
it is possible to edit the grob and have the calculations updated (see the
bottom-right panel of Figure 7.20).

```
> grid.edit("splitText", gp=gpar(cex=1.5))
```

The other way to encapsulate calculations with drawing operations is to use
the `grid.record()` function, as shown by the following code.

`drawDetails()` methods to replay on a device resize, it was necessary to turn off the recording of grid operations on the graphics engine display list, `engine.display.list(FALSE)`,
which forced grid to do all redrawing.

*Prior to R version 2.1.0, the call to `grid.text()` would need to add `recording=FALSE`.

```
1 splitTextGrob <- function(text, ...) {
2   grob(text=text, cl="splitText", ...)
3 }

5 drawDetails.splitText <- function(x, recording) {
6   grid.text(splitString(x$text),
7             x=0, y=1, just=c("left", "top"))
8 }
```

Figure 7.21
A "splitText" class. The `drawDetails` method for the class recalculates where to place line breaks in the text, based on the current viewport size.

```
> grid.record({
              grid.text(splitString(text),
                        x=0, y=1, just=c("left", "top"))
              },
              list(text=text))
```

This is convenient for writing code purely for its side-effect (i.e., without having to deal explicitly with grobs), but it provides less control over the design of the object that is created. There is also a `recordGrob()` function that simply creates a grob encapsulating the calculations and drawing operations without drawing anything.

Avoiding argument explosion

Very complex or high-level graphics functions and objects are usually composed of several lower-level elements, which in turn may be composed of several even-lower-level elements. For example, a scatterplot matrix is composed of several scatterplots and each scatterplot contains axes, labels, and data symbols.

Ideally, it should be possible to control any aspect of a graphical scene. In terms of writing code, this means that an argument or component should be supplied to allow the user to specify a customized value for any parameter of the scene.

At the level of graphical primitives, parameters consist of such things as the locations of lines, the color of lines, and the line thickness. At a higher level, for example for axes, there are higher-level parameters, such as where to place tick-marks, but it is also desirable to still be able to control the individual elements of the axis.

It is tempting to simply provide arguments for the elements of an axis as arguments of the axis itself. An example is where an an axis could have a `rot` argument to specify the angle of rotation of the tick-mark labels, but this approach quickly runs into difficulties. For one thing, ambiguities can easily arise. If an axis had an overall label it is unclear whether the `rot` argument would apply to the tick-mark labels or to the overall label. Another problem is that as elements become more complex, the number of parameters required for all sub-elements grows alarmingly. Consider the number of separate arguments required to individually specify the angle of rotation for tick mark labels on all scatterplots within a scatterplot matrix!

Grid provides several features that can help to solve this problem. The functions `grid.edit()` and `editGrob()` (see Section 6.1) make it possible to access the lower-level elements of an object using a gPath. For example, in the following code, an x-axis is created and then the labels on the tick marks are rotated by editing the `rot` component of the `text` grob called `"labels"` that is a child of the `xaxis` grob.

```
> grid.xaxis(at=1:3/4, name="xaxis1")
> grid.edit("xaxis1::labels", rot=45)
```

In the case where a grob calculates its children on the fly, it is possible to specify editing actions that should be applied on the fly. This typically occurs when a grob has no permanent children to access via a gPath and this will usually correspond to a grob that has a `drawDetails()` method.

The functions `gEdit()` and `gEditList()` allow the user to specify one or more edit operations and the functions `applyEdit()` and `applyEdits()` apply those operations to a grob. The following code demonstrates their use.* In this case, an x-axis is created without specifying the `at` argument, which means that tick-marks are calculated on the fly. The `edits` argument is used to specify modifications to the labels that will be generated on the fly.

```
> grid.xaxis(name="xaxis1")
> grid.edit("xaxis1", edits=gEdit("labels", rot=45))
```

This approach is similar to the concept of panel functions (see Sections 3.4.9 and 4.5).

---

* The `edits` component is not available for `xaxis` or `yaxis` grobs prior to R version 2.1.0, so this example will not run in earlier versions.

Mixing graphical functions and graphical objects

This chapter has addressed two main ways in which to develop new graphical functionality: as a graphics function (purely for the side-effect of producing output); and as a graphical object. There has also been an emphasis on producing reusable graphical elements, a corollary of which is that existing graphical elements should be used where possible in the construction of new graphical elements.

There is no way to force other developers to create graphical objects rather than graphical functions, so it is necessary to be able to make use of both existing functions and existing objects whether constructing a new function or a new object.

In order to discuss each of the four possible situations (new functions from existing functions, new functions from existing grobs, new grobs from existing functions, and new grobs from existing grobs) the following paragraphs consider the simple case of drawing a "face," which consists of a rectangle for the border, two circles for eyes, and a line for the mouth (see Figure 7.22 for examples).

Defining a new graphics function is straightforward whether using existing graphics functions or existing graphical objects. Figure 7.23 defines two new graphical functions to draw a face. The function `faceA()` demonstrates the most straightforward case of a graphics function that includes calls to other graphics functions to produce output. The function `faceB()` shows a graphics function making use of existing graphical objects, which is done by just passing the result of the object constructor function to the function `grid.draw()`.

Developing a new graphical object can be a bit trickier, but there are several tools to help out. Figure 7.24 defines three functions for creating a new graphical object to represent a face. The function `faceC()` shows the simplest case, where a gTree is built from existing graphical objects, by just creating the appropriate objects as children of the gTree.

The functions `faceD()` and `faceE()` demonstrate the harder problem of creating a new graphical object using only existing graphics functions. In the case of `faceD()`, the output of the graphics functions is captured as a gTree using the `grid.grabExpr()` function. The `faceE()` function shows a different approach: it creates a grob with a special class, `"face"`, and wraps the existing graphics functions in a `drawDetails()` method for that new class.[*]

---

[*]The function `grid.grabExpr()` is not available in R before version 2.1.0 so `faceD()` will not run in earlier versions. Also, the `drawDetails()` method for the `"face"` class that `faceE()` produces would need to have `recording=FALSE` added to each function call in order to run correctly.

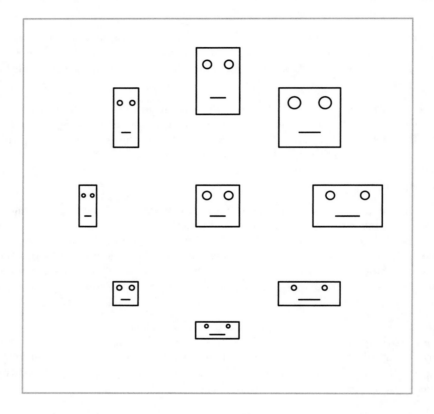

Figure 7.22
Drawing faces. Examples of the output that could be produced using the graphics
functions and graphical objects defined in Figures 7.23 and 7.24.

```
1 faceA <- function(x, y, width, height) {
2   pushViewport(viewport(x=x, y=y,
3                          width=width, height=height))
4   grid.rect()
5   grid.circle(x=c(0.25, 0.75), y=0.75, r=0.1)
6   grid.lines(x=c(0.33, 0.67), y=0.25)
7   popViewport()
8 }

10 faceB <- function(x, y, width, height) {
11   pushViewport(viewport(x=x, y=y,
12                          width=width, height=height))
13   grid.draw(rectGrob())
14   grid.draw(circleGrob(x=c(0.25, 0.75), y=0.75, r=0.1))
15   grid.draw(linesGrob(x=c(0.33, 0.67), y=0.25))
16   popViewport()
17 }
```

Figure 7.23
Some face functions. Some different ways to implement a new graphics function to draw a "face." The function faceA() makes use of existing graphics functions. The function faceB() makes use of existing graphical objects.

```
 1 faceC <- function(x, y, width, height) {
 2   gTree(childrenvp=viewport(x=x, y=y,
 3                             width=width, height=height,
 4                             name="face"),
 5         children=gList(rectGrob(vp="face"),
 6                        circleGrob(x=c(0.25, 0.75),
 7                                   y=0.75, r=0.1, vp="face"),
 8                        linesGrob(x=c(0.33, 0.67), y=0.25,
 9                                  vp="face")))
10 }

12 faceD <- function(x, y, width, height) {
13   grid.grabExpr({
14               pushViewport(viewport(x=x, y=y,
15                                     width=size,
16                                     height=size))
17             grid.rect()
18             grid.circle(x=c(0.25, 0.75),
19                         y=0.75, r=0.1)
20             grid.lines(x=c(0.33, 0.67), y=0.25)
21             popViewport()
22           })
23 }

25 drawDetails.face <- function(x, recording) {
26   pushViewport(viewport(x=x$x, y=x$y,
27                         width=x$width, height=x$height))
28   grid.rect()
29   grid.circle(x=c(0.25, 0.75), y=0.75, r=0.1)
30   grid.lines(x=c(0.33, 0.67), y=0.25)
31   popViewport()
32 }

34 faceE <- function(x, y, width, height) {
35   grob(x=x, y=y, width=width, height=height, cl="face")
36 }
```

Figure 7.24
Some face objects. Some different ways to implement a new graphical object to represent a "face." The function faceC() makes use of existing graphical objects. The function faceD() makes use of existing graphics functions by capturing their output as a gTree. The function faceE() makes use of existing graphics functions by creating a new class of grob with a special drawDetails() method.

7.4 Querying grid

This section describes some functions that are useful for querying the current state of the grid graphics system, which is useful for investigating the structure of the output that is produced by a function and for understanding what has gone wrong when a function has produced unexpected output (see also Section A.3.1 for some of the general debugging tools that R provides).

There are a number of functions that print out information about the current viewports and grobs in the current scene. The functions `current.viewport()` and `current.vpTree()` display information about the current viewport tree, the `getNames()` function lists the names of all top-level grobs in the current scene, and the function `childNames()` lists the names of all children of a gTree.

It can be particularly frustrating if an error results in output being placed in a strange location or, worse, no output being drawn at all. In such cases, it can be helpful to simply call the `grid.rect()` function, which will draw a rectangle showing the extent of the current viewport. Another useful trick is to calculate the location of the current viewport in inches, as shown by the following code; either a negative value or a very large value is usually a sign of trouble.

```
> convertX(unit(0:1, "npc"), "inches")
```

Chapter summary

It is possible to write simple grid graphics functions for the purpose of producing graphical output. Such functions should not assume that they have the entire device to draw into. They should only assume that they are drawing within a grid viewport. Naming any viewports created in the function and using `upViewport()` rather than `popViewport()` makes it possible for others to annotate the graphical output produced by the function. Naming all grobs produced by the function makes it possible for others to edit the output from the function (or remove grobs or add grobs or extract grobs).

Creating a graphical object, either a grob or a gTree, to represent the output generated by the function requires extra effort to set up methods for the new graphical object class, but provides additional benefits. Most graphical objects will be gTrees consisting of a high-level description plus several child grobs representing the output produced. A gTree makes it possible for others to interact with the high-level description, while still being able to access the low-level element grobs. A grob can also be useful to provide information about the amount of space required to produce graphical output. Finally, a grob makes it possible for others to create higher-level gTrees with the grob as a child element.

A

A Brief Introduction to R

This appendix provides a very brief introduction to R. No attempt is made to cover all aspects of R. Instead, the focus is on providing enough knowledge to be able to make use of the graphics functions described in the main part of this book.

An important part of this appendix is the definition of some technical terms that are used throughout the rest of the book. Each of these terms will be written using *italics* the first time it is used.

A.1 Obtaining and installing R

The R software and documentation can be obtained from the R home page: http://www.r-project.org/. There are binary versions of R for Windows, Mac OS X, and the main flavors of Linux. R is also known to build from source on a wide range of platforms, including several 64-bit architectures.

Installation of binary distributions varies depending on the platform, but is straightforward for anyone familiar with the relevant platform's standard installation tools.

A.2 An environment for statistical computing and graphics

R is run in the normal fashion (e.g., select it from the "start" menu on Windows, or type R at a unix command line). The appearance of the R environment will vary, but for the purposes of this book, the important thing is that there is always an R *command line*, indicated by a prompt, which is usually the "greater than" symbol, >. The various platforms only differ in terms of how pretty the surrounding windows look.*

There are several ways to interact with R. The most simple, and the most fun, is simply to type *expressions* at the command line. When an expression is complete, type <return> or <enter>, and R will *evaluate* the expression and print a result. The following code provides an example of typing a simple mathematical expression.

```
> 1 + 1
```

```
[1] 2
```

As well as mathematical expressions, R understands expressions involving functions (*function calls*), comparisons, and special control flow statements (see Section A.3). An example of a function call is shown by the following code.

```
> abs(-4)
```

```
[1] 4
```

The function call consists of the function name, abs in this case, followed by one or more *arguments* within brackets. In this case, there is only one argument, the number -4. The function abs() calculates the absolute value of a number.

Arguments to functions may be specified in the form name=value, in which case the arguments can be supplied in any order. Furthermore, arguments can have default values, in which case the argument does not need to be specified at all in a call to the function. The following three examples demonstrate these features using the seq() function, which generates regular sequences of numbers. The first three arguments to this function are from, a number to start with, to, a number to end with, and by, an increment. The result of the call to seq() is identical in all three cases below — the seq() function returns the integers from 1 to 10 inclusive — but each function call specifies the arguments in a different way.

---

*There are several packages and projects, for example the Rcmdr package[21] , the JGR project (http://stats.math.uni-augsburg.de/JGR/), and the SciViews-R project (http://www.sciviews.org/SciViews-R/), which provide more sophisticated windows-and-dialogs front-ends to R, but this book focuses on producing graphics by typing expressions at the command line.

The first expression uses positional arguments (values are matched to arguments in the order they are specified, e.g., the first value, 1, is matched to the first argument, `from`).

```
> seq(1, 10)
```

```
[1]  1  2  3  4  5  6  7  8  9 10
```

The second expression uses named arguments, so values are explicitly matched to arguments and the order does not matter.

```
> seq(to=10, from=1)
```

```
[1]  1  2  3  4  5  6  7  8  9 10
```

Both of the expressions above also make use of the default value for the `by` argument, which is 1, as demonstrated by the third expression.

```
> seq(to=10, from=1, by=1)
```

```
[1]  1  2  3  4  5  6  7  8  9 10
```

Many of the functions in this book do not return a value, but are used for their *side effect*, which is usually the production of graphical output. For example, the following code produces a scatterplot similar to the one shown in Figure 1.1, but does not return any value.

```
> plot(pressure)
```

A.2.1   Batch processing

Instead of typing expressions one at a time at the command line, it is possible to save several expressions in a text file (using a text editor and remembering that a word processor is not a text editor) and have R evaluate the expressions all at once. A semi-colon (;) can be used to indicate the end of each expression, but this is not necessary — the standard approach is simply to start each expression on a new line in the text file. On the other hand, long expressions may extend over several lines. Either way, R will provide errors messages if something is wrong.

The expressions in a file are evaluated either by starting R and calling the function source() with the name of the text file as an argument, or by running R in batch mode from a command-line shell. For example, suppose that yourfile.R is the name of the text file containing your R expressions. To run the expressions from within an interactive R session, start up R and type the expression below.

```
> source("yourfile.R")
```

To run the expression using R in batch mode, type the following command within a Linux shell or a Windows DOS shell or the equivalent.

```
R CMD BATCH yourfile.R
```

This command will produce a file called yourfile.Rout which contains the output from running the R code in yourfile.R, plus any graphics files that your code creates.

A.2.2   Data types

Unsurprisingly, R recognizes a number when the user types one (see the code examples above). To specify a piece of text (also called a *string*), type it within either single or double quotes (double quotes are always used in this book). The following code shows the use of one of R's many string functions to change text from lowercase to uppercase.

```
> toupper("some text")
```

```
[1] "SOME TEXT"
```

R also recognizes *logical values* (also called boolean values), which are typed as TRUE and FALSE, and there is a special value, NA, which represents a missing or unknown value.

In addition to these basic types of data, R provides a number of *data structures* that allow multiple values to be specified as a single object. In fact, the most basic type of data in R is a *vector* (several values) rather than a *scalar* (single value). A single value in R is actually just a vector of length 1.

There are numerous functions for generating vectors (seq() is one example) and many operations and functions automatically work element-wise on vectors. For example, the following code produces a vector of integer values from 1 to 10, adds the value 5 to each element of the vector, then calculates the average of the values in the vector.

```
> seq(10, 1)
```

```
 [1] 10  9  8  7  6  5  4  3  2  1
```

```
> seq(10, 1) + 5
```

```
 [1] 15 14 13 12 11 10  9  8  7  6
```

```
> mean(seq(10, 1) + 5)
```

```
[1] 10.5
```

Many graphics facilities are "vectorised" so that multiple pieces of output can be produced by a single function call and a vector of graphical parameter settings can be specified to a function call (see examples on pages 154 and 170).

In addition to vectors of numbers, strings, and logical values, R recognizes vectors of *categorical* values, where the elements of the vector may only take one of a finite set of values, such as `"male"` or `"female"`. Such a vector is called a *factor* and the set of possible values are the *levels* of the factor. Factors are most often automatically generated when data is imported into R, but a factor may also be generated explicitly using the `factor()` function. Some graphics functions are designed to work with factors to produce plots that are appropriate for categorical data.

A.2.3   Variables

R allows values to be assigned to (symbolic) *variables*; in other words, a value can be associated with a name and that name can be used to represent the value in subsequent expressions. Here is the previous example, using variables to store intermediate results. The expression `10:1` is just a short-hand for `seq(10, 1)`.

```
> x <- 10:1
> y <- x + 5
> mean(y)
```

```
[1] 10.5
```

The expression `x <- 10:1` can be read as "the variable x is assigned the value `10:1`". The assignment operator, `<-`, is a less-than sign followed by a minus sign *without any space between them.*

A.2.4 Indexing

R provides very powerful indexing facilities for accessing subsets of data. The following code demonstrates how to extract a single element, in this case the fourth element, from the vector of values in the variable x.

```
> x[4]
```

```
[1] 7
```

The subset (the bit within the square brackets) can be a vector of values and logical values can be used as well, as demonstrated by the following code.

```
> x[4:7]
```

```
[1] 7 6 5 4
```

```
> x[x > 3 & x < 8]
```

```
[1] 7 6 5 4
```

A.2.5 Data structures

Vectors are the most basic storage format in R, but more complex and high-dimensional structures are also available.

R provides support for matrices and matrix operations. The functions matrix(), cbind() and rbind() are useful for constructing matrix objects. Some graphics functions that plot three or more variables will accept the data to plot as a matrix and matrices are also sometimes useful when specifying the layout of several plots on a page. Matrices can be sub-setted using an index containing two values separated by a comma. The following code shows some simple examples of creating and sub-setting a matrix. First, a matrix consisting of three rows and two columns is created (and assigned to the variable m). Next, the value in the third row and second column is extracted. Finally, the entire second row is extracted.

```
> m <- cbind(1:3, 11:13)
> m
```

```
        [,1] [,2]
[1,]     1    11
[2,]     2    12
[3,]     3    13
```

```
> m[3, 2]
```

```
[1] 13
```

```
> m[2,]
```

```
[1]   2 12
```

All of the values in a vector or matrix must have the same basic type (e.g., they must all be numbers, or all be strings, or all be logical values). R provides *lists* to allow the user to store information of different types together in a single object. The following code shows how to construct a simple list and it also shows that names can be associated with components of the list.

```
> list("a", TRUE, c=1:3)
```

```
[[1]]
[1] "a"

[[2]]
[1] TRUE

$c
[1] 1 2 3
```

Each component of a list can be any type of structure, including a list, so lists provide a way of creating an arbitrarily complex collection of data. Indexing applies to a list just as for vectors to provide a sub-list, and there is an additional operation using double square brackets ([[) to provide the *contents* of an individual component of a list. The following code shows some simple examples. First, a list is created and assigned to the variable z, then we extract a sub-list, consisting of the first two components of z.

```
> z <- list("a", TRUE, c=1:3)
> z[1:2]
```

```
[[1]]
[1] "a"

[[2]]
[1] TRUE
```

Finally, the following code extracts just the contents of the third component
of z, which is a vector (using single square brackets would have produced a
list of length 1).

```
> z[[3]]
```

```
[1] 1 2 3
```

The subset can be specified as the names of components of the list and there
is a short-hand equivalent of the double square brackets using a dollar sign
($). The last example above could also be written as z[["c"]] or z$c.

R also supports *data frames* for storing a complete set of data (a set of variables
observed on a number of cases). This is the standard way of storing an entire
data set in R and a number of graphics functions allow the user to specify the
data to plot as a data frame. Data frames can be indexed like matrices or like
lists (where each variable is a component of the data frame) as shown by the
following code.

```
> df <- data.frame(x=1:3, y=factor(c("M", "F", "M")))
> df$x
```

```
[1] 1 2 3
```

```
> df[[2]]
```

```
[1] M F M
Levels: F M
```

```
> df[2, 1]
```

```
[1] 2
```

A.2.6 Formulae

R provides a special notation for specifying formulae that define statistical models. The expression y ˜ x indicates that the variable y is (linearly) dependent on the variable x. Within a formula, a plus symbol (+) is used to include further additive terms, a multiplication symbol (*) is used to indicate crossing of factors, a colon (:) is used to indicate an interaction, a forward slash (/) is used to indicate nesting, and a vertical bar (|) is used to indicate grouping.

Some traditional plotting functions allow a formula to express which variables to plot. For example, a scatterplot of the variables x and y can be obtained via the following expression.

```
> plot(y ~ x)
```

The Trellis graphics system, described in Chapter 4, makes extensive use of formulae to express the structure of a plot (see Section 4.2.1).

A.2.7 Expressions

Usually, when an expression is typed at the R command line, two things happen: R *parses* what has been typed to create a valid R expression and then R *evaluates* the expression. Sometimes it is useful to do just the first step and generate an unevaluated expression (see Section 3.4.6). There are several ways to do this, but the most straightforward is to use the **expression()** function. The following code shows the difference between getting R to evaluate an expression and getting R to create an expression.

```
> x <- 10:1
> y <- 5

> x + y

 [1] 15 14 13 12 11 10  9  8  7  6

> expression(x + y)

expression(x + y)
```

A.2.8   Packages

R functions are organized into separate *packages*, which has two benefits.
Firstly, this makes it possible to load only the packages containing the functions that are required, so that R runs faster and uses less memory. Secondly,
it is very easy to make use of functions that other people have written and
put into a package (the main R download site currently contains more than
400 of these "contributed" packages).

The functions `install.packages()` and `update.packages()` are useful for
downloading and installing packages from within an R session (the Windows
version has a menu interface to these). Once a package is installed, it can be
loaded (so that the functions in the package can be used) using the `library()`
function.

The traditional graphics system described in Part I of this book is provided by
the `graphics` package, which is installed and loaded by default in a standard
installation of R. The grid graphics system is provided by the package `grid`.
It and the lattice package described in Part II are also installed by default,
but must be explicitly loaded. See `help(Startup)` for ways to automate the
loading of packages.

A.2.9   Accessing data sets

R has many facilities for obtaining data. There are a number of datasets
provided with R in the `datasets` package. Typing `data()` will produce a list of
the datasets available in all currently loaded packages. For importing external
data sets into R, there are several functions for reading plain text files in a
wide variety of formats (e.g. `read.table()`, `read.csv()`, and `read.fwf()`).
The package `foreign`[49] allows the user to read in data that has been stored
in another system's format (e.g., SAS data sets), and there are functions for
working with database management systems (see the `DBI` package[50]).

It can also be useful to generate random data, for example, when learning
how to use a function. For this purpose, there are many functions in R for
producing random values from a wide variety of distributions. For example,
the `rnorm()` function generates random values from a Normal distribution.

The manual "R Data Import/Export", which is distributed with R, contains
a comprehensive description of facilities for accessing data.

A.2.10 Getting help

Every R function and dataset has on-line help associated with it, which can be viewed by calling the `help()` function. Try typing `help(help)`. The help pages often include a series of examples and the `example()` function is useful for running those examples. The `help.start()` function starts up a browser to view HTML versions of the on-line help, plus HTML versions of the R manuals. There is also a `help.search()` function to assist in locating a function for a particular purpose.

Some packages also provide *vignettes*, which are usually more extensive and descriptive than the on-line help. Typing `vignette()` will show a list of the vignettes that are available for all currently loaded packages.

Finally, there are a number of manuals distributed with R. As well, there are freely available manuals contributed by users on the R web site (including several in languages other than English), and there are archives of the `R-help` mailing list (also linked off the main web site).

If all else fails, try sending a nice clear question to the `R-help` mailing list (see the instructions and posting guide for "Mailing lists" on the R home page).

A.3 A programming language

One of the most important features of R is that there is a deliberate blurring of the distinction between users and developers — between people who call functions and people who write functions.

It is very easy to write your own function in R, and such a function can be used just like any predefined function. The following code gives an example of how to define a function. This function takes a vector and extends it by attaching a mirror image of the vector to the original vector.

```
> mirror <- function(x) {
          c(x, rev(x)[-1])
          }
```

It is now possible to call this function just like any other function, as shown by the following code. The result of a function is just the result of the last expression in the function (but, see also the `return()` function).

```
> mirror(1:4)
```

```
[1] 1 2 3 4 3 2 1
```

There are many details to add to this basic recipe (see the manual "An Introduction to R" that is distributed with R), but creating simple functions is as simple as it looks.

As more sophisticated functions are written, some of the programming-related features of R become more important. R provides the standard control flow features, such as loops and conditional expressions. The following code redefines the `mirror()` function above, using a conditional expression to add a check that it is possible to mirror the value passed in.

```
> mirror <- function(x) {
            if (!is.vector(x))
               stop("x must be a vector")
            c(x, rev(x)[-1])
          }
```

The call to the `stop()` function will only occur if an inappropriate value is passed in the argument x, as shown by the following code.

```
> mirror(1:4)
```

```
[1] 1 2 3 4 3 2 1
```

```
> mirror(matrix(1:10, ncol=2))
```

```
Error in mirror(matrix(1:10, ncol = 2)) :
    x must be a vector
```

A.3.1 Debugging

When writing your own functions, or even just when calling existing functions, it can sometimes be unclear why a particular error has occurred, or even where it has occurred. R provides a number of tools for investigating the behavior of a function more closely, including basic functions like `str()` that display the raw contents of variables, the function `traceback()`, which shows all the functions that were called on the way to getting an error, and the `debug()` function, which allows the user to step through a function one command at a time. There is also a `debug` package[9], which provides a GUI debugger interface.

A.4 An object-oriented language

R provides some support for object-oriented concepts such as *classes* and *generic functions.** All objects in R have a class, which can be determined using the `class()` function, as shown by the following code.

```
> x <- TRUE
> class(x)
```

```
[1] "logical"
```

Some functions in R are generic, which means that the behavior of the function depends on the class of the first argument to the function. For a generic function, there can be a number of different *methods* where each method is a function that corresponds to the action to be taken for a particular class. For example, the `plot()` function is generic and there are different methods for drawing a different sort of plot when the first argument, x, is a factor compared to when the first argument is numeric (just some numbers).

It is simple to define a new method for a generic function. The name of the method function must be of the form `generic.class`, where `generic` is the name of the generic function and `class` is the class that the new method is for. The following code demonstrates how to define a new method for the generic `plot()` function, which will be called when the first argument, x, is a logical vector. This method just prints out a crude character-based plot consisting of a "bar" of Ts representing the number of `TRUE` values in the argument and a "bar" of Fs for the `FALSE` values.

```
> plot.logical <- function(x, ...) {
    cat(paste(rep("T", sum(x)), sep="", collapse=""),
        "\n",
        paste(rep("F", length(x) - sum(x)),
            sep="", collapse=""),
        "\n",
        sep="")
}
```

---

*There are actually two object-oriented systems in R. The one used in the graphics systems described in this book is much more basic and is referred to as the S3 object system. There is also a fully-featured object system provided by the `methods` package, which is referred to as the S4 object system, but a discussion of that is beyond the scope of this book.

This method will now get run whenever `plot()` is called with a logical vector as the first argument, as shown by the following code.

```
> plot(1:10 > 4)
```

TTTTTT
FFFF

These object-oriented features are important in the development of new graphical objects for the grid graphics system.

B

Combining Traditional Graphics and Grid Graphics

The grid graphics system and the traditional graphics system work completely independently of each other. However, with care, the output from the two systems can be combined effectively.*

B.1 The gridBase package

The grid graphics system offers more power and flexibility than the traditional graphics system, and the lattice package provides the standard plot types and some facilities not available in the traditional system. However, it is often necessary to use the traditional system because most plotting functions in add-on packages for R are built on the traditional system. Clearly, a combination of the wide range of traditional plots and the power and flexibility of grid and lattice would be desirable and this is what the gridBase package[44] provides.

B.1.1 Annotating base graphics using grid

The gridBase package has one function, baseViewports(), that supports adding grid output to a base graphics plot. This function creates a set of grid viewports that correspond to the current base plot. By pushing these viewports, it is possible to do simple annotations to a traditional plot, such as adding lines and text using grid's units to locate them relative to a wide variety of coordinate systems, or to attempt more complex annotations involving pushing further grid viewports.

The baseViewports() function returns a list of three grid viewports. The first corresponds to the base "inner" region. This viewport is relative to the entire

---

*This appendix is derived from material that first appeared as an article in R News[44].

device and it only makes sense to push this viewport from the "top level" (i.e., only when no other grid viewports have been pushed). The second viewport corresponds to the base "figure" region and is relative to the inner region, and it only makes sense to push it after the "inner" viewport has been pushed. The third viewport corresponds to the base "plot" region and is relative to the figure region, and it only makes sense to push it after the other two viewports have been pushed in the correct order.

A simple application of this facility involves adding text to the margins of a base plot at an arbitrary orientation. The base function `mtext()` allows text to be located in terms of a number of lines away from the plot region, but only at rotations of 0 or 90 degrees. The base `text()` function allows arbitrary rotations, but only locates text relative to the user coordinate system in effect in the plot region (which is inconvenient for locating text in the margins of the plot). By contrast, the grid function `grid.text()` allows arbitrary rotations and can be used in any grid viewport. In the following, a base plot is created with the x-axis tick labels left off.

```
> midpts <- barplot(1:10, col="grey90", axes=FALSE)
> axis(2)
> axis(1, at=midpts, labels=FALSE)
```

Next, `baseViewports()` is used to create grid viewports that correspond to the base plot and those viewports are pushed.

```
> library(gridBase)
> vps <- baseViewports()
> pushViewport(vps$inner, vps$figure, vps$plot)
```

Finally, rotated labels are drawn using `grid.text()` (and the viewports are popped to clean up). The final output is shown in Figure B.1.

```
> grid.text(c("one", "two", "three", "four", "five",
              "six", "seven", "eight", "nine", "ten"),
           x=unit(midpts, "native"), y=unit(-1, "lines"),
           just="right", rot=60)
> popViewport(3)
```

B.1.2   Embedding base graphics plots in grid viewports

The gridBase package provides several functions for adding base graphics output to grid output. There are three functions that allow base plotting regions

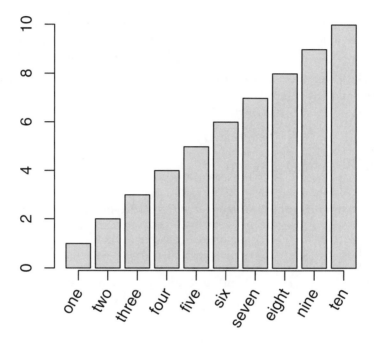

Figure B.1

Annotating a traditional plot with grid. Most of the plot is drawn using the traditional `barplot()` function, but the x-axis labels are drawn using `grid.text()` to make use of both a convenient coordinate system (lines of text away from the x-axis) and the ability to rotate text to any angle.

to be aligned with the current grid viewport. These make it possible to draw one or more base graphics plots within a grid viewport. The fourth function, gridPAR(), provides a set of graphical parameter settings so that base par() settings can be made to correspond to some* of the current grid graphical parameter settings.

The first three functions are gridOMI(), gridFIG(), and gridPLT(). They return the appropriate par() values for setting the base "inner," "figure," and "plot" regions, respectively.

The main usefulness of these functions is to allow the user to create a complex layout using grid and then draw a base plot within relevant elements of that layout. The following example uses this idea to create a lattice plot where the panels contain dendrograms drawn using base graphics functions.†

The first step just involves preparing some data to plot. A dendrogram object is created and cut it into four subtrees.‡

```
> hc <- hclust(dist(USArrests), "ave")
> dend1 <- as.dendrogram(hc)
> dend2 <- cut(dend1, h=70)
```

Next, some dummy variables are created that correspond to the four subtrees.

```
> x <- 1:4
> y <- 1:4
> height <- factor(round(unlist(lapply(dend2$lower,
                                        attr, "height")))) 
```

Now a lattice panel function is defined to draw the dendrograms. The first thing this panel function does is push a viewport that is smaller than the viewport lattice creates for the panel. The purpose of this is to ensure that there is enough room for the labels on the dendrogram. The space variable contains a measure of the length of the longest label. The panel function then calls gridPLT() and makes the base plot region correspond to the viewport that has just been pushed. Finally, the traditional plot() function is used to draw the dendrogram (and then the viewport is popped).

---

*Only lwd, lty, col were available at the time of writing.

†Recall that lattice is built on grid so the panel region in a lattice plot is a grid viewport.

‡This example uses data on violent crimes in the United States, available as the USArrests data set in the datasets package.

```
> space <- 1.2 * max(unit(rep(1, 50), "strwidth",
                           as.list(rownames(USArrests)))))
> dendpanel <- function(x, y, subscripts, ...) {
    pushViewport(viewport(gp=gpar(fontsize=8)),
                 viewport(y=unit(0.95, "npc"), width=0.9,
                          height=unit(0.95, "npc") - space,
                          just="top"))
    par(plt=gridPLT(), new=TRUE, ps=8)
    plot(dend2$lower[[subscripts]], axes=FALSE)
    popViewport(2)
  }
```

The final plot is produced by a call to the xyplot() function, using lattice to set up the arrangement of panels and strips (grid viewports) and our panel function to draw a traditional dendrogram in each panel (see Figure B.2).

```
> library(lattice)
> plot.new()
> print(xyplot(y ~ x | height, subscripts=TRUE,
               xlab="", ylab="",
               strip=function(...) {
                 strip.default(style=4, ...)
               },
               scales=list(draw=FALSE),
               panel=dendpanel),
        newpage=FALSE)
```

B.1.3 Problems and limitations

The functions provided by the gridBase package allow the user to mix output from two quite different graphics systems and there are limits to how much the systems can be combined:

- The gridBase functions attempt to match grid graphics settings with base graphics settings (and vice versa). This is only possible under certain conditions. For a start, it is only possible if the device size does not change. If these functions are used to draw into a window, then the window is resized, the base and grid settings will almost certainly no longer match and the graph will become nonsensical. This also applies to copying output between devices of different sizes.

  Some solutions to this problem are discussed in Section 7.3.10.

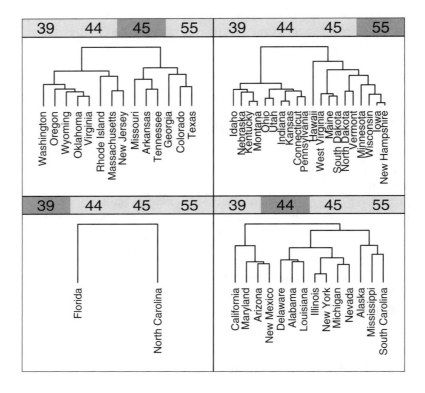

Figure B.2
Embedding a traditional plot within lattice output. The arrangement of the panels
and the drawing of axes and strips is all done by lattice using grid, but the contents
of each panel is a dendrogram plot produced by the traditional graphics system.

- It is not possible to embed base graphics output within a grid viewport that is rotated.

- There are certain base graphics functions that modify settings like `omi` and `fig` themselves (e.g., `coplot()`). Output from these functions will not embed properly within grid viewports.

Bibliography

[1] A contribution to computer typesetting techniques: Tables of coordinates for Hershey's repertory of occidental type fonts and graphic symbols. *NBS Special Publication 424*, April 1976.

[2] Daniel Adler. *rgl: 3D visualization device system (OpenGL)*, 2004. R package version 0.64-13.

[3] Adobe Systems Inc. *PostScript language reference manual (2nd ed.)*. Addison-Wesley Longman, 1990.

[4] Jens Henrik Badsberg. *dynamicGraph: dynamicGraph*, 2004. R package version 0.1.7.0.

[5] Richard A. Becker and John M. Chambers. *Extending the S System*. Chapman & Hall, 1985.

[6] Richard A. Becker, William S. Cleveland, and Ming-Jen Shyu. The visual design and control of trellis display. *Journal of Computational and Graphical Statistics*, 5:123–155, 1996.

[7] Richard A. Becker, Allan R. Wilks, and R version by Ray Brownrigg with enhancements by Thomas P. Minka. *maps: Draw Geographical Maps*, 2005. R package version 2.0-26.

[8] Roger Bivand, Friedrich Leisch, and Martin Mächler. *pixmap: Bitmap Images ("Pixel Maps")*, 2004. R package version 0.4-2.

[9] Mark V. Bravington. *MVB's debugger for R*, 2004. R package version 1.0.1.

[10] Dan Carr and R port by Nicholas Lewin-Koh and Martin Maechler. *hexbin: Hexagonal Binning Routines*, 2004. R package version 1.1.

[11] J. M. Chambers. Structured computational graphics for data analysis. *Proceedings of the International Statistical Institute*, 40, 1975.

[12] William S. Cleveland. *The Elements of Graphing Data*. Wadsworth Publ. Co., 1985.

[13] William S. Cleveland. *Visualizing Data*. Hobart Press, 1993.

[14] William S. Cleveland and Marylyn E. McGill, editors. *Dynamic Graphics for Statistics*. Wadsworth, 1988. ISBN 0-534-09144-X.

[15] William S. Cleveland and Robert McGill. Graphical perception: The visual decoding of quantitative information on graphical displays of data (C/R: p210-229). *Journal of the Royal Statistical Society, Series A, General*, 150:192–210, 1987.

[16] A. Cohen. On the graphical display of the significant components in a two-way contingency table. *Communications in Statistics — Theory and Methods*, A9:1025–1041, 1980.

[17] D. Cook, A. Buja, J. Cabrera, and C. Hurley. Grand tour and projection pursuit. *Journal of Computational and Graphical Statistics*, 4:155–172, 1995.

[18] Peter Dalgaard. *Introductory Statistics with R*. Springer, 2002. ISBN 0-387-95475-9.

[19] M.W. Felgate, S.H. Bickler, and P. Murrell. Quantifying broken objects: Estimating the death assemblage by integrating sample assemblage brokenness and completeness. *Journal of Archaeological Science*, submitted.

[20] John Fox. *An R and S-Plus Companion to Applied Regression*. Sage Publications, Thousand Oaks, CA, USA, 2002. ISBN 0761922792.

[21] John Fox. *Rcmdr: R Commander*, 2005. R package version 0.9-17.

[22] M. Friendly. Graphical methods for categorical data. *SAS User Group International Conference Proceedings*, 17:190–200, 1992.

[23] M. Friendly. A fourfold display for 2 by 2 by k tables. Technical Report 217, Psychology Department, York University, 1994.

[24] M. Friendly. Mosaic displays for multi-way contingency tables. *Journal of the American Statistical Association*, 89:190–200, 1994.

[25] Michael Friendly. *Visualizing Categorical Data*. SAS Publishing, 2000.

[26] Frank E Harrell, Jr. *Hmisc: Harrell Miscellaneous*, 2004. R package version 3.0-1.

[27] Mark A. Harrower and Cynthia A. Brewer. Colorbrewer.org: An online tool for selecting color schemes for maps. *The Cartographic Journal*, 40:27–37, 2003.

[28] J.A. Hartigan and B. Kleiner. A mosaic of television ratings. *The American Statistician*, 38:32–35, 1984.

[29] Richard Heiberger and Burt Holland. *Statistical Analysis and Data Display: An Intermediate Course with Examples in S-PLUS, R, and SAS*. Springer, 2004.

[30] Torsten Hothorn, Kurt Hornik, and Achim Zeileis. Unbiased recursive partitioning: A conditional inference framework. Research Report Series 8, Department of Statistics and Mathematics, WU Wien, 2004.

[31] Ross Ihaka. *colorspace: Colorspace Manipulation*, 2005. R package version 0.9.

[32] L. Kaufman and P.J. Rousseeuw. *Finding Groups in Data: An Introduction to Cluster Analysis*. Wiley, New York, 1990.

[33] Duncan Temple Lang and Debby Swayne. *The embedded ggobi R interface*, 2001. R package version 0.3.

[34] H. W. Lie and B. Bos. *Cascading Style Sheets, level 1*, 1996. W3C Recommendation.

[35] Uwe Ligges and Martin Mächler. Scatterplot3d — an R package for visualizing multivariate data. *Journal of Statistical Software*, 8(11):1–20, 2003.

[36] Ulric Lund and R port by Claudio Agostinelli. *CircStats: Circular Statistics*, 2003. R package version 0.1-8.

[37] John Maindonald and John Braun. *Data Analysis and Graphics using R: an example-based approach*. Cambridge University Press, 2003.

[38] S. McClatchie and T.M. Ward. Alongshore variation in upwelling intensity in the eastern Great Australian Bight. *Journal of Geophysical Research*, in press.

[39] Doug McIlroy and packaged for R by Ray Brownrigg and Thomas P. Minka. *mapproj: Map Projections*, 2004. R package version 1.1-7.

[40] D. R. McNeil. *Interactive Data Analysis*. Wiley, 1977.

[41] David Meyer, Achim Zeileis, Alexandros Karatzoglou, and Kurt Hornik. *vcd: Visualizing Categorical Data*, 2004. R package version 0.1-3.4.

[42] A.E. Miller. The analysis of unreplicated factorial experiments from a geometric perspective. *Canadian Journal of Statistics*, 31:311–327, 2003.

[43] P. R. Murrell. Layouts: A mechanism for arranging plots on a page. *Journal of Computational and Graphical Statistics*, 8:121–134, 1999.

[44] Paul Murrell. Integrating grid graphics output with base graphics output. *R News*, 3(2):7–12, October 2003.

[45] Paul Murrell and Ross Ihaka. An approach to providing mathematical annotation in plots. *Journal of Computational and Graphical Statistics*, 9:582–599, 2000.

[46] Kurt Nassau, editor. *Color for Science, Art and Technology*. Elsevier, 1998. ISBN 0-444-89846-8.

[47] Erich Neuwirth. *RColorBrewer: ColorBrewer palettes*, 2004. R package version 0.2-2.

[48] Martyn Plummer, Rolando Herrero, Silvia Franceschi, Chris J.L.M. Meijer, Peter Snijders, F. Xavier Bosch, Silvia de Sanjosé, and Nubia Muñoz for the IARC Multi-centre Cervical Cancer Study Group. Smoking and cervical cancer: pooled analysis of the iarc multi-centric case-control study. *Cancer Causes and Control*, 14:805–814, 2003.

[49] R Core Team, Saikat DebRoy, Roger Bivand, et al. *foreign: Read Data Stored by Minitab, S, SAS, SPSS, Stata, Systat, dBase, ...*, 2004. R package version 0.8-4.

[50] R Special Interest Group on Databases (R-SIG-DB). *DBI: R/S-Plus Database Interface*, 2003. R package version 0.1-8.

[51] Naomi Robbins. *Creating More Effective Graphs*. Wiley, 2005.

[52] Peter Rousseeuw, Anja Struyf, Mia Hubert, R port by Kurt Hornik, and Martin Mächler. *cluster: Functions for clustering (by Rousseeuw et al.)*, 2004. R package version 1.9.7.

[53] P.J. Rousseeuw. A visual display for hierarchical classification. In E. Diday, Y. Escoufier, L. Lebart, J. Pages, Y. Schektman, and R. Tomassone, editors, *Data Analysis and Informatics 4*, pages 743–748. North-Holland, Amsterdam, 1986.

[54] Deepayan Sarkar. Lattice. *R News*, 2(2):19–23, June 2002.

[55] Dave Schreiner. *OpenGL Reference Manual: The Official Reference Document to OpenGL, Version 1.2*. Addison-Wesley Longman, 1999.

[56] A. Struyf, M. Hubert, and P.J. Rousseeuw. Integrating robust clustering techniques in S-PLUS. *Computational Statistics and Data Analysis*, 26:17–37, 1997.

[57] Deborah F. Swayne, Duncan Temple Lang, Andreas Buja, and Dianne Cook. GGobi: evolving from XGobi into an extensible framework for interactive data visualization. *Computational Statistics and Data Analysis*, 43:423–444, 2003.

[58] Jürgen Symanzik. Interactive and dynamic graphics. In James E. Gentle, Wolfgang Härdle, and Yuichi Mori, editors, *Handbook of Computational Statistics*. Springer, 2004.

[59] Terry M Therneau, Beth Atkinson, and R port by Brian Ripley. *rpart: Recursive Partitioning*, 2004. R package version 3.1-21.

[60] E. R. Tufte. *The Visual Display of Quantitative Information*. Graphics Press, 1989.

[61] Edward R. Tufte. *Envisioning Information*. Graphics Press, 1990.

[62] Simon Urbanek. *iPlots - interactive graphics for R*. R package version 0.1-18.

[63] Bill Venables and R port by Kurt Hornik. *oz: Plot the Australian coastline and states.* R package version 1.0-10.

[64] William N. Venables and Brian D. Ripley. *Modern Applied Statistics with S. Fourth Edition.* Springer, 2002. ISBN 0-387-95457-0.

[65] John Verzani. *Using R for Introductory Statistics.* Chapman & Hall/CRC, Boca Raton, FL, 2005. ISBN 1-584-88450-9.

[66] Denis White. *maptree: Mapping, pruning, and graphing tree models,* 2003. R package version 1.3-3.

[67] Leland Wilkinson. *The Grammar of Graphics.* Springer, 1999.

Function Index

base package
debug, 276
expression, 97
is.finite, 120
is.na, 120
jitter, 41
load, 217
on.exit, 119
read.csv, 274
read.fwf, 274
read.table, 274
save, 217
source, 217
substitute, 97
traceback, 276

CircStats package
rose.diag, 5

cluster package
plot.agnes, 29

gplots package
plotCI, 18

graphics package
abline, 87
arrows, 87
assocplot, 41
axis, 59, 94–96, 117, 158
axis.Date, 96
axis.POSIXct, 96
axTicks, 96
barplot, 27, 32, 34, 106, 108
box, 73, 88
boxplot, 27, 32, 108, 119
boxplot.stats, 119
bxp, 104

chull, 86
close.screen, 82
co.intervals, 119
contour, 35, 41, 103, 119
contourLines, 119
coplot, 38, 77, 110, 119
curve, 29
dotchart, 38
erase.screen, 82
filled.contour, 35, 94, 110
fourfoldplot, 35
frame, 114
grid, 86
hist, 27, 119
identify, 41
image, 35, 41
layout, 77–83, 109
layout.show, 78
lcm, 80
legend, 76, 92, 101
lines, 27, 83–85, 90, 112, 117
locator, 41
matlines, 86
matplot, 29, 86
matpoints, 86
mosaicplot, 38
mtext, 55, 89–90, 117
n2mfrow, 118
nclass.FD, 119
nclass.scott, 119
nclass.Sturges, 119
pairs, 38, 109, 110
panel.smooth, 110
par, 48–78, 99, 119
persp, 35, 112
pie, 27
plot, 1, 27–29, 34–35, 55

plot.new, 114
plot.window, 115
plot.xy, 115
points, 27, 55, 83, 90
polygon, 86
rect, 86, 117
rgb, 56
rug, 88
screen, 82
segments, 85, 117
split.screen, 82, 83
stars, 5, 38
stem, 29
strheight, 99
stripchart, 29
strwidth, 99, 117
sunflowerplot, 41
symbols, 35, 103
text, 64, 85, 112
title, 90
xfig, 26
xinch, 101
xy.coords, 118
xyinch, 101
xyz.coords, 118
yinch, 101

grDevices package
bitmap, 20
bmp, 20
cm, 101
cm.colors, 57
col2rgb, 56
colorRamp, 58
colorRampPalette, 58
colors, 55
colours, 55
convertColor, 56
dev.control, 22, 254
dev.copy, 21
dev.copy2eps, 21
dev.cur, 20
dev.list, 20
dev.next, 21
dev.off, 19

dev.prev, 21
dev.print, 21
dev.set, 20
dev2bitmap, 21
getGraphicsEvent, 41, 191
gnome, 20
graphics.off, 21
gray, 57
grey, 57
hcl, 56
heat.colors, 57
hsv, 56
jpeg, 20
palette, 56
pdf, 20
pictex, 20
png, 20
postscript, 20
postscriptFont, 65
postscriptFonts, 65
quartz, 20
rainbow, 57
recordGraphics, 101
rgb, 55
rgb2hsv, 56
terrain.colors, 57
topo.colors, 57
win.metafile, 20
windows, 20
X11, 20
x11, 20
xfig, 20

gridBase package
baseViewports, 279
gridFIG, 282
gridOMI, 282
gridPAR, 282
gridPLT, 282

grid package
addGrob, 201
applyEdit, 258
applyEdits, 258
arrowsGrob, 155

childNames, 204, 263

circleGrob, 155

convertHeight, 162

convertUnit, 162

convertWidth, 162, 254

convertX, 162

convertY, 162

current.viewport, 263

current.vpTree, 181, 263

dataViewport, 151

downViewport, 152, 177, 182, 236

drawDetails, 163, 236–241, 254

editDetails, 241–245

editGrob, 201, 234, 241, 258

engine.display.list, 253

frameGrob, 211, 213

gEdit, 258

gEditList, 258

get.gpar, 168

getGrob, 201

getNames, 203, 263

gList, 203

gpar, 166

gPath, 205

grid.add, 201

grid.arrows, 155, 156, 190, 197

grid.circle, 155

grid.display.list, 253

grid.draw, 207, 236

grid.edit, 152, 200–203, 234, 241, 253, 258

grid.frame, 211

grid.get, 200, 201

grid.grab, 209, 217

grid.grabExpr, 210

grid.layout, 185–189

grid.line.to, 154, 155, 179, 190

grid.lines, 155

grid.locator, 191

grid.move.to, 154, 155, 179, 197

grid.newpage, 150

grid.pack, 211

grid.place, 212

grid.points, 155

grid.polygon, 155, 156, 190

grid.prompt, 151

grid.record, 163, 256

grid.rect, 155, 263

grid.refresh, 253

grid.remove, 201

grid.segments, 155

grid.set, 201

grid.text, 155

grid.xaxis, 155, 157

grid.yaxis, 155, 157

grob, 231, 232

grobHeight, 164, 214

grobWidth, 164, 214

gTree, 208, 231, 232

heightDetails, 245–246

linesGrob, 155

lineToGrob, 155, 197

moveToGrob, 155

packGrob, 213

placeGrob, 213

plotViewport, 151

pointsGrob, 155

polygonGrob, 155

popViewport, 176

postDrawDetails, 246–248

preDrawDetails, 246–248

pushViewport, 174

recordGrob, 257

rectGrob, 155

removeGrob, 201

seekViewport, 178, 197

segmentsGrob, 155

setGrob, 201

stringHeight, 164

stringWidth, 164

textGrob, 155

unit, 159–166

unit.c, 162

unit.length, 162

unit.pmax, 162

unit.pmin, 162

unit.rep, 162

upViewport, 152, 177, 236

validDetails, 234–236

viewport, 173–189

vpList, 181
vpPath, 182
vpStack, 181
vpTree, 181
widthDetails, 245–246
xaxisGrob, 155
yaxisGrob, 155

gtkDevice package
devGTK, 20

hexbin package
hexbin, 41

Hmisc package
errbar, 18
labcurve, 32

lattice package
barchart, 131
bwplot, 131, 133, 147
canonical.theme, 139
cloud, 131, 133
col.whitebg, 139
contourplot, 131
densityplot, 131
dotplot, 131, 147
equal.count, 135
histogram, 131
levelplot, 131
llines, 147
lpoints, 147
ltext, 144
panel.xyplot, 144
parallel, 131
print.trellis, 191
qq, 131
qqmath, 131
shingle, 135
show.settings, 137
splom, 131
stripplot, 131
trellis.device, 128
trellis.focus, 147, 194
trellis.identify, 147

trellis.panelArgs, 147
trellis.par.get, 137
trellis.par.set, 139
trellis.vpname, 197
update, 128
wireframe, 131, 133
xyplot, 131

maps package
map, 110

pixmap package
addlogo, 106

RJavaDevice package
devJava, 20

RSvgDevice package
devSVG, 20

stats package
lm, 29
na.omit, 120
plot.dendrogram, 41
plot.lm, 29

utils package
example, 26
help, 25
help.start, 25
str, 276
vignette, 149

vcd package
ternaryplot, 5

Concept Index

3D plots, 3, 35, 131
  annotating, 112–114

Annotating plots
  in grid, 184
  lattice plots, 142–147, 193–194
  traditional plots, 83–114
    3D plots, 112–114
    plot margins, 89–92
    plot region, 83–88
Appearance of output,
  see Graphical parameters
Arguments, 266–267
  recycling, 161, 170
  standard
    in grid, 158–159
    in lattice plots, 130
    in traditional plots, 34–35
Arranging plots, *see also* Layouts
  lattice plots, 140–142
  traditional plots, 77–83
Arrows, *see* Graphical primitives
Assignment, 269
Association plots, 41
Axes
  in grid, 151–152, 157
  in lattice plots, 135, 136, 142
  in traditional plots, 34, 70–73
    as annotation, 94–96
    color, 55
    date-based, 96
    font, 65
    line style, 59

Bannerplots, 29
Barcharts, 131
Barplots, 3, 27

  annotating, 106–108
  customizing, 32–34
Batch processing, 267–268
Bitmap
  adding to a plot, 106
  output format, 20
Boolean values, *see* Logical values
Box-and-whisker plot, *see* Boxplot
Boxplots, 3, 27, 131
  annotating, 108–109
  customizing, 32–34

Categorical data, 269
  plotting, 5, 38, 41
Charts, *see* Plots
Circles, *see* Graphical primitives
Classes, 277–278,
  see also Generic functions
  defining new, 232–234
Clip art, 6
Clipping
  in traditional plots, 76
  viewports, 179
Color
  device dependence, 58
  in grid, 167
  semitransparent, 56–57
  specifying, 55–56
  in traditional plots, 55–59
Color sets, 57–58
Color spaces, 55–56
Command line, 265
Conditioning plots, 38,
  see also Multipanel conditioning
  annotating, 110
Contour plots, 35
Control flow, 276

Conversions,
   see Coordinate systems
Coordinate systems
   converting between, 99–101, 162–163
   in grid, 150, 152, 159–162,
     see also Units
   normalized, 48, 146, 160
   polar, 5
   in traditional plots, 47–48, 99–101
   user coordinates, 47
CRAN, vii
Cross-hatching, 58
Curves, 29, 158

Data frame, 27, 133, 272
Data sets
   accessing, 274
Data structures, 270–272,
   see also Classes
Data symbols, 68–70, 83, 155
Data types, 268–269
Debugging, 276
Dendrograms, 29, 41, 282
Density plots, 131
Devices, *see* Graphics devices
Display list
   in graphics engine, 21–22
   in grid, 200–201, 253–257
     searching, 206–207
Documentation, *see* Help
Dotplots, 38, 131
Drawing context, 173
Dynamic graphics,
   see Interactive graphics

Editing output, 200–203, 217–220
Ellipsis argument, 34, 119
Embedding output
   grid in lattice, 193–194
   grid in traditional, 279–280
   lattice in grid, 194–198
   traditional in grid, 280–283
Equation, mathematical,
   see Mathematical formulae

Error bars, 17
Expressions, 97, 266, 273

Factors, 269
Figure margins, 44
   annotating, 89
Figure region, 44
   clipping, 76
Fill patterns, 58–59
Flow of control, *see* Control flow
Formulae, 273
   as data to plot, 27, 86, 133–135
   in text,
     see Mathematical formulae
Fourfold display, 35
Frames, 210–213
   packing grobs, 211–212
   placing grobs, 212–213
Functions, 266,
   see also Graphics functions
   generic, *see* Generic functions
   writing, 275–276

Generic functions, 277
   in grid, 231–232, 249
   traditional, 27, 86
Geometric context, 173
gLists, 203
gPaths, 205
Graphical context, 173
   editing, 216
   explicit, 168
   implicit, 168
Graphical objects
   developing new, 231–259
     standard components, 234
     standard methods, 249
   drawing, 207, 236–241
   editing, 201, 241–245
   grouping, 208
   list of, *see* gList
   off-screen, 207–210
   path, *see* gPath
   pre- and post-drawing, 246–248
   reusing, 251

sizing, 245–246
tree of, *see* gTree
validating, 234–236
Graphical parameters, 166–172,
  see also Graphics state
editing, 216
in grobs, 158
in gTrees, 205–206
specifying, 169–170
vectorized, 170–172
in viewports, 184
Graphical primitives
in grid, 154–158
in traditional plots, 83–86
Graphical utilities, 86–88
Graphics device packages, 16,
  see also Packages
Graphics devices, 16, 20–21
Graphics engine, 16
Graphics file formats,
  see Output formats
Graphics functions
types of, 17–18
writing,
  see Writing graphics functions
Graphics object,
  see Graphical objects
Graphics packages, 16,
  see also Packages
Graphics state, 48–52,
  see also Graphical parameters
Graphics systems, 16
combining, 279
comparison of, 18
Graphs, *see* Plots
Grobs, *see* Graphical objects
Grouping output,
  see Graphical objects
gTrees, 203–206, 208–209

Help, 275
Hershey outline fonts, 65
special font faces, 66, 170
High-level functions, 17
Histograms, 3, 27, 131

Indexing, 270–272
Inner region, 44
Interactive graphics, 41–42, 191,
  see also Editing output

Jittering data, 41, 108

Keys, 142, 249, *see also* Legend
Key strokes, 42, 191

Layouts
  in grid, 185–189
  in traditional plots, 78–82
  nested, 187–189
Legends, 4, 92–94, 246,
  see also Keys
Level plots, 131
Line plots, 27, 84, 102
Lines,
  see also Graphical primitives
  end and join style, 60
  type of, 59–60
  width of, 59
Lists, 271–272
Locales, 68
Logarithmic transformations, 73
Logical values, 268
Low-level functions, 17

Maps, 5, 110, 225–227
Mathematical formulae
  in text, 97–99
Matrices, 270–271
  plotting, 29, 86
Methods, 277–278,
  see also Generic functions
Missing values, 268
  in grid, 190–191
  in traditional plots, 88–89
Modifying output,
  see Editing output
Modularity, 223
Mosiacplots, 38
Mouse events, 42, 191
Mulivariate plots,

see Plots, of multiple variables
Multipanel conditioning, 133

Non-finite values
    in grid, 190–191
    in traditional plots, 62, 88–89
Normalised coordinates,
    see Coordinate systems
Numeric values, 277

Object-oriented features, 277–278
Objects,
    see Classes and Graphical objects
Outer margins, 44
    annotating, 89
Output formats, 20
Overlapping output,
    see Overlaying output
Overlaying output, 101–106,
    see also painters model

Packages, 274
    base, 265–278
    CircStats, 5
    cluster, 29
    colorspace, 56
    datasets, 274
    DBI, 274
    debug, 276
    dynamicGraph, 42
    foreign, 274
    gplots, 18
    grDevices, 19–22
    gridBase, 279–285
    gtkDevice, 20
    hexbin, 17
    Hmisc, 18, 32
    iPlots, 42
    mapproj, 5
    maps, 5, 17, 110
    maptree, 41
    methods, 277
    oz, 225
    pixmap, 106
    Rcmdr, 266

RColorBrewer, 58
    Rggobi, 38, 42
    rgl, 38
    RGraphics, ix
    RJavaDevice, 20
    rpart, 41
    RSvgDevice, 20
    scatterplot3d, 38
    tcltk, 42
    vcd, 5, 41
Packing, *see* Frames
Painters model, 3, 150
Panel functions
    in lattice, 144–146
    in traditional plots, 109–110
par, *see* Graphics state
Parallel coordinate plots, 131
Parameters, *see* Arguments
Parameters, graphical,
    see Graphical parameters
Physical size, *see also* Units
    accuracy of, 100
    of graphics device, 20
    in layouts, 80
    of line width, 59
    of text, 64
    of viewports, 150
Piecharts, 3, 27
Placing, *see* Frames
Plot region, 44
    annotating, 83
    clipping, 76
Plots
    annotating, *see* Annotating plots
    arranging, *see* Arranging plots
    modern, 38–41, 130
    of multiple variables, 35–38
    specialized, 38–41, 130
    standard types, 27, 130
Plotting regions, 44
Plotting symbols,
    see Data symbols
Polar coordinates,
    see Coordinate systems
Polygons, *see* Graphical primitives

Prepanel function, 144
Primitives, *see* Graphical primitives
Programming, 275,
    see also Writing graphics functions

Quantile plots, 131

Rectangles, *see* Graphical primitives
Recycling, *see* Arguments, recycling
Regular expressions, 201
Rug plots, 88

S-Plus, vii–viii, 193
Saving output, 217
Scatterplot matrix, 38, 131
Scatterplots, 3, 27, 131
Shingles, 135
Star plots, 38
Stem-and-leaf plots, 29
Strip function, 144
Stripplots, 131
Sub-setting, *see* Indexing
Sunflower plots, 41
Superimposing output,
    see Overlaying output

Ternary plots, 5
Text, *see also* graphical primitives
    encoding, 65, 68
    font, 65–68
    justification of, 62–64
    multi-line, 65
    rotation, 64
    size of, 64
Three-dimensional plots,
    see 3D plots
Transformations,
    see Coordinate systems
Transformations, logarithmic,
    see Logarithmic transformations
Trellis graphics, 126

Units, 159–162
    absolute versus relative, 230
    in layouts, 185–187
    operations on, 161–162

    to size and position output, 162
    to size and position viewports, 173
User coordinates,
    see Coordinate systems

Variables, 269
Vectors, 268–269
Viewport lists, 181
Viewport paths, 182–183
Viewports, 173–184
    clipping, *see* Clipping
    drawing between, 179
    in grobs, 183–184
    in gTrees, 206
    navigating between, 177–178
    nesting, 175
    popping, 176–177
    pushing, 174–176
Viewport stacks, 181
Viewport trees, 181–182
    current, 181
Vignettes, 275
    for grid, 149

Writing graphics functions
    for traditional plots, 118–120
    in grid, 223–230